The Mechanisms of Explosions

The Mechanisms of Explosions

27 Case Studies for their Understanding

Jacques Chaineaux

iSTE

WILEY

First published 2023 in Great Britain and the United States by ISTE Ltd and John Wiley & Sons, Inc.

First published in French as “Les explosions accidentelles : 27 cas pour comprendre les mécanismes d’une explosion” by CNPP Editions © 2019.

ISTE Ltd
27-37 St George’s Road
London SW19 4EU
UK

www.iste.co.uk

John Wiley & Sons, Inc.
111 River Street
Hoboken, NJ 07030
USA

www.wiley.com

Library of Congress Control Number: 2022946011

British Library Cataloguing-in-Publication Data
A CIP record for this book is available from the British Library
ISBN 978-1-78630-886-3

Contents

Foreword

Lessons learned is the start of preventing accidents with gases and other substances or materials, mixed with or without air, that can explode and cause serious injury, even fatalities and significant damage to assets and the environment. Explosions all produce a "bang" but the mechanisms of how they occur can be very different. Knowledge of the mechanism is essential for the prevention of an explosion. In his book *The Mechanisms of Explosions – 27 Case Studies for Their Understanding*, Jacques Chaineaux reports the important feedback from the numerous technical investigations and judicial expertise cases that he has carried out in this field during 40 years of his professional life. In each case, he proposes recommendations that would have avoided the accidents discussed in this book.

In Part 1, he introduces the main concepts needed to characterize the explosion phenomenon and proposes a systematic method for conducting investigations and drafting explosion expert reports.

Then, he describes the successive steps to be followed to determine the mechanism of an explosion. First, the author focuses on identifying the damage to people and property caused by its mechanical and thermal effects. The reader is then led to ask themselves three questions: how much, what and how? The answers lead the reader to evaluate the magnitude of the energy released by the explosion, the quantity of the explosive system involved, its nature and finally the conditions that enabled its formation, in coherence with its circumstances.

In Part 2, he describes in detail the mechanism of 27 representative case studies of accidental explosions, selected from industrial as well as domestic ones, to cover a number of physical or chemical phenomena, substances and mechanisms, without limiting them to common situations but also very specific ones. The comprehensive and relevant lessons learned from each case constitute a real reference guide, the use of which can be instructive to envisage and explain other scenarios of potential accidents and to prevent them effectively.

Certainly, this work allows an approach of the expertise in the field of explosion to a broad audience of engineers, security managers, judicial and insurance experts, technical staff of local authorities, trainers as well as students of universities and colleges. It is therefore very commendable that Jacques has taken the effort to consolidate his experiences and knowledge in this book, and I hope that many readers will benefit from it.

Hans PASMAN
College Station, Texas, USA, September 16, 2022

Professor Dr. Ir. Hans J. Pasman is Emeritus Professor of Chemical Risk Management at the Delft University (NL) and currently Research Professor at Process Safety Center at Texas A&M University. After his PhD, he joined the Dutch organization TNO to perform research in reactive materials, explosions, investigations of industrial accidents and risk analysis. Then he was the former Chairman of the European Federation of Chemical Engineering (EFCE) Working Party on Loss Prevention and Safety Promotion in the Process Industries and co-founder of the European Process Safety Center.

Acknowledgments

In the introduction to this book, I indicate that my motivation for writing this book was the desire to pass on the experience I have gained in analyzing explosions. This desire is nothing new, but it has grown over the years. The first people who have benefited from this experience were my young colleagues, whom I accompanied at the end of my professional career: when a new investigation arrived, it was them whom I asked to lead it with me, and the first step was often to discover the site of the explosion together.

While writing, I often thought of them and I want to thank them here because they often told me that they appreciated my willingness to share my experience with them, which encouraged me to synthesize it into a book.

In 2020, I published a French version of this book entitled *Les explosions accidentelles – 27 cas pour comprendre les mécanismes d'une explosion* and I greatly appreciated my collaboration with the CNPP editorial board (www.cnpp.com). I especially want to thank them because I could not have published this English version if they had not given me back their rights.

Finally, my warmest thanks go to Hans Pasman, who agreed to write the Foreword to this book.

Jacques Chaineaux

Introduction

An explosion is characterized by the release of a large amount of energy, in an extremely short amount of time. The occurrence of this very brief phenomenon involves a process (a mechanism), the beginning and the end of which can be defined. As a result of the release of this energy, two types of effects are produced, mechanical ones, by gas expansion (air pressure wave and debris projection), and possibly thermal ones.

It is essential to determine the phenomenon or the sequence of phenomena involved in an explosion, with a triple objective:

– to identify all the elements of the cause of the explosion and to establish its mechanism, based on its circumstances and the effects it has produced;

– to define appropriate measures to prevent the occurrence of a new explosion and also to protect against the effects of those that would occur, despite the measures taken to prevent them;

– to generalize these measures by eventually translating them into regulatory constraints, recommendations or good practices.

I call "investigation of accidental explosions" the work required to achieve the first objective; the results obtained thus increase knowledge of the situations that may lead to the occurrence of explosions, as well as of the various possible mechanisms. Achieving the second objective is a logical follow-up to the first one and this work is one part of the more comprehensive activity of feedback (REX) of accidents. Finally, the mechanism of certain explosions has a generality character, which must be taken into account in order to achieve the third objective.

I practiced the activity of "investigation of accidental explosions" for almost 40 years, first at CERCHAR (*Centre d'études et recherches de Charbonnages de*

France) and then at INERIS (*Institut national de l'environnement industriel et des risques*), and this led me to work on a few hundred cases. In this book, I report on this investigation activity by detailing 27 case studies, chosen to describe the broadest spectrum of possible mechanisms.

I propose a classification of these mechanisms, according to the different physical phenomena involved, the different substances involved in each phenomenon and the initial conditions of each explosion (circumstances, confined or unconfined explosion, observed effects, etc.). Each case was chosen by the concern to treat at least one explosion within each class of mechanism and by emphasizing the generality of each case as well as its particularities.

It should be noted that there are many accidental explosions in which some elements of the cause are the result of human error, and the prevention of human errors is of course an important part of explosion prevention. In some of the cases presented, human errors are identified as real elements of cause, but are not considered in detail.

The risk of explosion is present in any industrial activity, since experience shows that accidental explosions occur in a wide range of industrial types, during the operation of different types of installations and as a result of the implementation of various hazardous substances in different processes which may give rise to an explosion under certain conditions. In the domestic environment, explosions are almost exclusively due to gas.

The term "accidental explosions" reflects the fact that there are also situations in which "voluntary" explosions are produced under conditions that are both sought-after and controlled (use of explosives, so-called "thermal" engines, fireworks, etc.).

In addition to causing human damage by killing or injuring people who are subject to their effects more or less severely, explosions also cause damage of a mechanical or thermal nature, the severity of which varies depending on the nature and quantity of the substances used. An explosion at an industrial facility can cause such serious economic damage (loss of operations, repair/reconstruction costs) that it can even jeopardize the survival of the plant, or even the activity at the industrial site (this was the case for the explosion of the AZF site in Toulouse in 2001). In the domestic environment, a traditional construction is weak since it is not able to sustain the level of overpressure that an explosion develops and it is common for the destruction to be significant and the persons present in the destroyed premises to be fatally or seriously injured.

On the whole, the objective of this book is to pass on my experience and to help anybody who is called upon to practice this activity of investigation. This book is divided into two parts:

– Part 1 provides general information about the explosion phenomenon, including simple theoretical aspects, necessary for understanding the classification of the different mechanisms identified (see Chapter 1), before moving on to present the general approach that I propose: to carry out the investigation of an accidental explosion (see Chapter 2).

– Part 2 then presents each of the 27 case studies.

PART 1

General Information and Approach

1

The Explosion Phenomenon

Any explosion involves one of the following four explosive systems which meet the definition given in the introduction and for which the explosion involves either a chemical reaction (systems 1 and 2) or a physical phenomenon (systems 3 and 4):

1) the explosive system is an Explosive Atmosphere (ATEX);

2) the explosive system is a substance or a mixture of substances which may give rise to a highly exothermic chemical reaction, which may be explosive under certain conditions;

3) the explosive system is a hollow body that breaks down under pressure of its contents, which is in the gaseous state and possibly also in the liquid state;

4) the explosive system consists of a hot body that comes into contact with a cold liquid and causes its vaporization very fast.

Each explosive system is characterized by the energy E_{lib}, which is released during the explosion. The higher the E_{lib} value, the greater the intensity of the explosion effects.

In the next section, for each explosive system, the elements defining the explosive phenomenon involved and an assessment of the value of E_{lib} are presented successively.

1.1. Explosion of an ATEX

1.1.1. *Definition of an ATEX*

An ATEX is the mixture with air, in some proportions, of a “combustible” or “flammable” product, which may be:

– a gas;

– a vapor emitted from a liquid;

– a liquid dispersed as an aerosol;

– or a dust dispersed as a cloud.

Whatever the state of the product, the explosion of an ATEX always results from the combustion reaction of that product in the oxygen of the air.

1.1.1.1. *Vocabulary elements*

When the product is a gas, a vapor or an aerosol, the explosion is referred to as a “gas explosion”, while when the combustible product is a dust, it is referred to as a “dust explosion”.

The combustion results in a “flame” which is the “chemical reactor” where the combustion reaction occurs, since it is in the flame where very hot fumes (over 1,000 K) are produced. These fumes are mainly flue gases, and it is these gases that undergo a thermal expansion and are responsible for the effects of the explosion.

The terms commonly used are “flammable gas” or “flammable liquid” (although it is the vapor emitted by the liquid that is flammable, since the combustion takes place in the gaseous phase). For a dust, the term is “combustible dust” instead of “flammable dust”. However, these expressions are equivalent, since any flammable or combustible product undergoes combustion with the propagation of a flame. Sometimes, in the following, the word “fuel” has been used as a synonym of “flammable gas or liquid”.

Once the combustion reaction is initiated, the flame propagates to the whole ATEX volume. However, this propagation can only occur if the content of the combustible product in ATEX is between two limits, the lower explosion limit or LEL and the upper explosion limit or UEL.

The European standards, as NF-EN 1839 (NF-EN 1839 2012), give the definitions of the LEL and UEL, as well as of the explosion range:

– the explosion range is the range of content of a flammable product in mixture with air or with a premixture of air and inert gas, within which an explosion can occur;

– the LEL and UEL are the lower and upper limit of the explosion range, determined under specified conditions.

Finally, the wordings "lower flammability" or LFL and "upper flammability limit" or UFL are used in some English documents and in US standards. Numerically, they can be considered as equal to LEL and UEL.

1.1.1.2. *Ignition energy*

The combustion of an ATEX does not occur spontaneously, since for it to start, the ATEX must receive a certain amount of energy locally, which is referred to as "ignition energy". There is a minimum value of this energy, which is referred to as "minimum ignition energy" (MIE), so that ignition occurs only if the energy supplied to ATEX is at least equal to the MIE.

MIE is likely to be released by different ignition sources and those most frequently involved in accidental explosions are related to each of the following phenomena:

– a naked flame, such as a torch, a burner or a lighter;

– the phenomenon of arc or electric spark (short circuit or electrical circuit break);

– the phenomenon of electrostatic arc or spark;

– the phenomenon of incandescent friction particles (e.g. grinding or cutting of steel);

– the phenomenon of self-ignition, which occurs when the ATEX is heated to a temperature greater than its self-ignition temperature (AIT), i.e. the minimum temperature from which ATEX ignites spontaneously, without any energy input other than that provided by its own temperature.

1.1.1.3. *Similarities and differences between an ATEX explosion and a fire*

In this book devoted to explosions, there will be several opportunities to consider the phenomenon of fire, especially when dealing with domestic accidents that have involved both an explosion and a fire.

These two phenomena present both similarities and differences. Before discussing the detailed study of the ATEX explosion phenomenon, it is useful to compile (as seen in Table 1.1) the main similarities and differences between the fire and ATEX explosion phenomena.

Type of phenomenon	**Fire**	**Explosion of an ATEX**
Type of chemical reaction	Chemical reaction of exothermic combustion of a combustible product in the air	
Fuel and air mixture	Combustible and air mixture pre-exists in the form of an ATEX. Flame propagation is only possible if the combustible and air contents are in the explosion range	The combustible is heated by the fire; if it is a solid (a liquid, respectively), it gradually produces pyrolysis gases (vapors, respectively) that mix with the air and burn immediately
Ignition	Ignition energy must be brought locally to ATEX. MIE depends on the composition of ATEX and equals a fraction of mJ at stoichiometry for gases and vapors	Ignition energy varies widely depending on the nature and volatility of the combustible. A spark of a few mJ is enough to start the fire of a volatile liquid (gasoline), but it takes more than 1,000 J to ignite paper and even more to ignite a compact solid (wood)
Flame temperature	The temperature of a flame propagating in an ATEX depends on the composition of the ATEX. For a stoichiometric composition, it can reach 3,000 K	The temperature of a fire is lower than the flame temperature of an ATEX, as flammable gases or vapors are always in excess of the air. This excess results in the production of smoke, containing more or less soot
Kinetic data	The combustion speed is governed by the speed at which the flame propagates in the ATEX (from a few tens of cm/s to a few m/s)	The combustion speed is governed by the combustible production speed, which is significantly slower than the combustion speed of an ATEX. Fire smoke is very hot and can help spread the fire
Moving from one phenomenon to the other	The flame of an explosion can start a fire of a combustible material when it is in contact with it or is exposed to its radiation	The flame of a fire contains sufficient energy to be a possible source of ignition of an ATEX

Table 1.1. *Similarities and differences between the fire and ATEX explosion phenomena*

1.1.1.4. *Case of explosion of an ATEX consisting of a flammable gas*

There are many flammable gases: hydrogen, carbon monoxide CO, hydrocarbons such as alkanes (methane, propane, butane, etc.), alkenes (ethylene, propylene, etc.), alkynes such as acetylene, etc.

The MIE of most flammable gases is less than 1 mJ. The values of the LEL and UEL are available in the literature (INRS ED 911 2004). The "explosive range" is the content range between the LEL and UEL, and its amplitude varies greatly depending on the nature of the gas. For example:

– for hydrogen, the range is very wide, between 4 and 75% volume;

– for n-butane, the range is much smaller, between 1.8 and 8.5 % volume;

In section 1.1.4.1, the combustion of an air–butane ATEX, as well as the energy it produces, according to its composition, are treated in detail.

1.1.1.5. *Case of explosion of an ATEX consisting of vapor emitted by a flammable liquid*

The conditions for the formation of an ATEX from a flammable liquid depend on both the liquid itself and its environment.

1.1.1.5.1. Conditions specific to the liquid itself

Existence of a flash point

Any flammable liquid is characterized by the existence of a flash point FP, which is defined by a standard (NF-EN 13237 2012) as the lowest temperature for which "the display of a flame causes the inflammation of vapors emitted by a sample of the liquid under prescribed test conditions".

Experimental determination of a liquid's flash point consists of regularly heating a few centiliters of liquid in a cup, then occasionally approaching the flame of a small gas burner to the surface of the liquid, until a small and fugitive flame appears as a "flash", i.e. a flame is visible and immediately dies.

There are different standards for the determination of the FP, in particular two main types of methods, which are:

– methods known as "open-cup";

– "closed-cup" methods.

For the same liquid, the flash point measured with a closed cup is always 1 to 3 degrees lower than the flash point measured with an open cup: this is because, when heating the sample in a closed cup, the vapors emitted by the liquid remain confined in the sky of the cup and do not disperse into the air as they do in the case of an open cup. As a result, the lowest temperature for which the air over the liquid is ignited is a slightly lower value for the closed cup than for the open cup.

The knowledge of a flash point is important because various regulations relating to flammable liquids are based on its value. Table 1.2 shows the flash point values of flammable liquids commonly used in the industry, from the lowest to the highest.

Liquid	Flash Point, FP (°C)	Liquid	Flash point, FP (°C)
Diethyl-ether	-45	Toluene	4
n-Pentane	-40	Methanol	11
n-Hexane	-22	Ethanol	12
Acetone	-20	1-Propanol	15
Cyclohexane	-20	n-Butyl-acetate	22
Benzene	-11	1-Butanol	29
Ethyl-acetate	-4	o-, m-, p-xylene	32, 27, 27

Table 1.2. *Flash points of common flammable liquids*

The existence of a flash point reflects the ability of a liquid to emit vapors which, mixed with air, can form an ATEX and is therefore the first characteristic to be taken into account when assessing the risk of forming an ATEX from a liquid.

Interference between the temperature of the liquid and FP

A liquid emits more vapors when it is brought to a higher temperature: as a result, the risk of formation of ATEX from a liquid depends not only on the existence of a flash point but also on the difference between the temperature T at which the liquid is applied and FP.

For an ATEX to be formed, at least the following inequality [1.1] must be satisfied:

$$(T - FP) \geq 0 \qquad [1.1]$$

However, the risk of ATEX formation from vapors emitted by a liquid does not depend only on the relative values of T and FP, but may also depend on the environment of the liquid.

1.1.1.5.2. Conditions relative to the environment of the liquid

The environmental characteristics of a liquid, which may affect the risk of formation of an ATEX from the vapors it emits, are relative to:

– the possibility that a liquid–vapor equilibrium could be established;

– the existence of an air flow that sweeps the surface of the liquid, which is sufficient to disperse the vapors emitted by the liquid.

Case of an existence of a liquid-vapor equilibrium

The risk of formation of ATEX from vapors emitted by a liquid depends not only on its flash point but also on the possibility that the thermodynamic equilibrium between the liquid and the vapor can be established, which implies that the following conditions are met simultaneously:

– the liquid is contained in a closed container;

– a liquid phase coexists with the vapor phase (the vapor is then called "saturated", whereas it is called "dry" in the absence of a liquid phase).

If a flammable liquid is contained in a closed container where a liquid–vapor equilibrium can be established, the risk of formation of an ATEX in the container is no longer strictly dependent on its flash point, but on a quantity called the "lower explosion point" (LEP).

Definition of LEP and UEP

The LEP of a liquid is defined in the standard (NF-EN 1839 2012) as the temperature at which vapors emitted by the liquid and in thermodynamic equilibrium with it, form, when mixed with air under atmospheric pressure, an ATEX, the composition of which is that of a mixture at the lower explosion limit (LEL).

Similarly, the UEP of a liquid may be defined as the temperature at which vapors emitted by the liquid and in thermodynamic equilibrium with it, form, when mixed with air under atmospheric pressure, an ATEX, the composition of which is that of a mixture at the upper explosion limit (UEL).

In Figure 1.1, the x-axis corresponds to the vapor content of a vapor–air mixture under atmospheric pressure and the y-axis corresponds to the temperature of that mixture. This diagram provides a graphical representation of the temperature influence on the LEL and UEL of the vapor emitted by a flammable liquid and enables the representative points of the LEP and UEP to be shown.

In the container, a liquid–vapor equilibrium is established and the P_{sat} pressure of the vapor in equilibrium with the liquid depends only on the temperature of the liquid. Once the equilibrium has been achieved, the atmosphere in the container contains a mixture of vapor and air, homogeneous throughout the container's sky, and the content of vapor C_v (expressed indifferently in percentage volume or percentage molar) is equal to:

$$C_v = 100\, P_{sat}/P_{atm} \qquad [1.2]$$

where P_{atm} is atmospheric pressure.

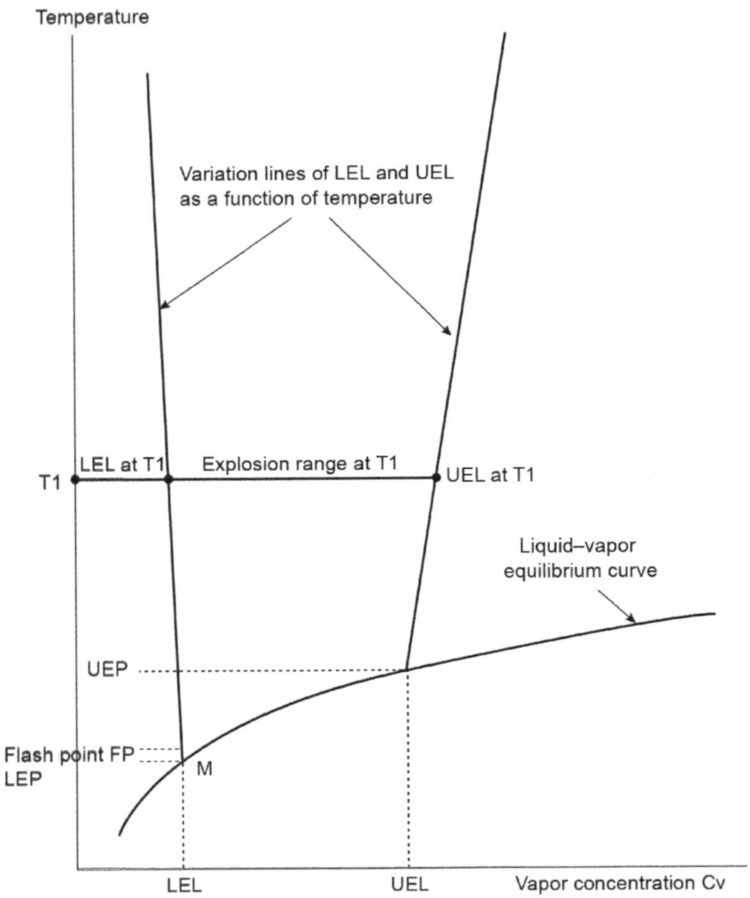

Figure 1.1. *Representation of the influence on the LEL and UEL of the vapor emitted by a flammable liquid, enabling the representative points of LEP, UEP and FP to be shown*

The atmosphere consisting of an air–vapor mixture is an ATEX if the C_v vapor content belongs to the explosion range delimited by the LEL and the UEL. The dependence of the LEL and UEL with a temperature higher than UEP is linear according to Burgess–Wheeler's law (USBM B503 1953), with a negative slope for the LEL and a positive slope for the UEL. Figure 1.1 shows the interference of the liquid–vapor equilibrium curve of a flammable liquid with the explosion range of the vapor emitted by that liquid:

– the combustion taking place in the gaseous phase: it is the vapor emitted by the liquid that is flammable and not the liquid itself;

– at any liquid–vapor equilibrium corresponds a value of the vapor saturation pressure emitted by the liquid P_{sat}. The ratio P_{sat}/P_{atm} corresponds to the C_v content of the vapor in this mixture, expressed as percentage volume, where C_v is defined according to the relationship [1.2];

– the temperature dependence of LEL with temperature cuts the liquid–vapor equilibrium curve at an M-point, the abscissa and ordinate of which are, respectively, the LEL and the LEP.

As noted above, the flash point of a flammable liquid is the lowest temperature for which the vapors emitted may be ignited under specified conditions, but these conditions do not correspond to those of a liquid–vapor equilibrium (the standards for determination of the flash point require a continuous heating of the liquid which prevents the establishment of a thermodynamic equilibrium).

The FP value is thus slightly higher than the LEP; the difference is 2 to 3°C for a pure liquid, but the difference can reach about 15°C for a mixture of liquids of different volatility (e.g. a petroleum distillation cut).

However, contrarily to FP, LEP represents the temperature threshold below which no ATEX can be formed under confinement conditions: in the temperature range between LEP and FP, it means that there is a risk of ATEX formation in all conditions where a liquid is contained in a container and a liquid–vapor equilibrium can be established.

1.1.1.5.3. Case of an absence of a liquid-vapor equilibrium

There are many industrial situations where no liquid–vapor equilibrium can be established, but where it is essential to characterize the risk of formation of an ATEX. For example, these are all situations where a flammable liquid is implemented in an open environment and where this liquid may be spread on the ground. Since no liquid–vapor equilibrium can be established, the formation of an ATEX thus depends on the FP of the liquid and not on its LEP.

If the inequality [1.1] is verified, it is possible to state that an ATEX is certainly present in the immediate vicinity of the free surface of the liquid. It is therefore important to be able to estimate as precisely as possible the volume and the position of this ATEX, because the intensity of the effects that its explosion could produce depends on these volume and position.

To evaluate the volume of ATEX formed by evaporation of a flammable liquid spread on the ground of a room, it is necessary to successively:

– firstly evaluate the evaporation rate q of this liquid;

– and secondly examine the conditions under which the vapors produced are mixed with ambient air and diluted at a rate that is dependent on the room's air renewal.

Estimated evaporation rate of a liquid puddle spread on the ground

INRS document ND 2313 (Chaineaux INRS ND 2313 2009), entitled "Conditions de formation d'une ATEX lors de la mise en œuvre d'un liquide inflammable" presents a model that was established on the basis of a test campaign involving three different flammable liquids, contained in a square container of 1 m^2 and subjected to an air flow of 0.3 or 0.9 m/s. The evaporation rate q(Ts) of each liquid can be determined by using the following relationship [1.3]:

$$q(Ts) = 22\ S.V.M^{0.4}(P_{sat}(Ts)/P_{atm}) \qquad [1.3]$$

where:

– q(Ts) is the liquid evaporation rate at Ts (in g/s);

– Ts is the temperature of the liquid at its surface (in K);

– S is the surface of the liquid puddle (in m^2);

– V is the air velocity at the surface of the puddle (in m/s);

– M is the molar mass of the liquid (in kg/mole);

– $P_{sat}(Ts)$ is the saturated vapor pressure of the liquid at Ts (in Pa);

– P_{atm} is the atmospheric pressure supposed to be 101,300 Pa.

The above relationship is applicable for the following domains:

– for Ts = Teb (Teb is the boiling temperature of the liquid under atmospheric pressure), the evaporation rate is no longer governed by Ts but also by the effect of boiling;

– S may be up to several m^2 for $0.3 < v < 0.9$ m/s (the flow q(Ts) is proportional to S and, beyond certain dimensions of the puddle, the air becomes saturated by the vapor before reaching the edge of the puddle and the proportionality is no longer respected); since the tests were performed for v = 0.3 and 0.9 m/s, the interpolation between these values is possible, but extrapolation is more uncertain, particularly for very high values of V;

– to express q(Ts) in kg/h, its value in g/s must be multiplied by the coefficient 3.6.

From the q(Ts) value, the following relationship [1.4] determines for what minimum value of the air flow Q_{min}, the air–vapor mixture formed downstream of the puddle is sufficiently diluted by air to be less than 25% LEL, and thus to ensure that no significant volume of ATEX is formed.

$$Q_{min} = (100\ q(Ts).RT/P_{atm})/(M.LEL) \qquad [1.4]$$

where:

– Q_{min} is the minimum air flow rate (in Nm^3/h);

– q(Ts) is the liquid evaporation rate (in kg/h);

– R is the gas constant = 8.31 J. K^{-1}. $mole^{-1}$;

– T is the air temperature (in K);

– P_{atm} is the atmospheric pressure supposed to be 101,300 Pa;

– M is the molar mass of the liquid (in kg/mole);

– LEL is the lower explosion limit of the liquid (in percentage volume).

As an example, for a puddle of n-hexane having a 1 m^2 surface (molar mass M = 86 g/mole, LEL = 1.1% volume), spread on the ground of a room in ambient conditions (Ts = 20°C, P_{atm}) for which the saturated vapor pressure is $P_{sat}(Ts)$ = 16 kPa, and being submitted to a horizontal 1 m/s air flow, the evaporation rate q(Ts) equals 4.7 kg/h and Q_{min} equals 480 m^3/h.

1.1.1.6. *Case of an ATEX containing an inert gas*

When an inert gas is added to an ATEX consisting of a mixture of air and any fuel, the explosion range of the fuel in the mixture of air and inert gas diminishes.

To understand the influence of an inert gas on the flammability of a fuel product in the presence of air, it is relevant to visualize the explosion domain of the fuel in a triangular diagram of the ternary air–fuel–inert system thus constituted.

In the case of an accidental explosion in which an air–fuel–inert flammable mixture is likely to have exploded, the use of a triangular diagram often enables the flammability of the mixture to be characterized by its composition, and also to determine the quantity of one of the constituents (the fuel or air) whose addition could make flammable the initial mixture.

The most common inert gases are N_2 and CO_2. Specifically, for any combustible product, as a gas, a vapor or a dust, its oxygen limit content (LOC) for a given inert gas is defined as the O_2 limit content, below which any ATEX cannot be formed, regardless of the fuel content.

As part of the investigation of an explosion involving a hydrogenated ATEX (see Part 2 case 7), a triangular representation of the ternary systems H_2–air–N_2 or H_2–air–CO_2 has been used, which shows that the area of the explosion domain of the second system is less than that of the first, which means that CO_2 is an inert gas more efficient than N_2.

More generally, it is common that the search for the mechanism and causes of an explosion require us to characterize the flammability of a mixture consisting of one or more flammable gases mixed with air (or even oxygen-enriched air) and one or more inert gases. In general, one of the following questions should be answered:

– Could the explosion have resulted from the ignition of such a defined composition mixture?

– What is the limit proportion of air (i.e. O_2 oxidizer) added to a mixture of an inert gas (or a defined mixture of several inert gases) and a flammable gas (or a defined mixture of several flammable gases) for which the resulting mixture reaches its explosion limit, whatever the proportions of the inert gas(es) and the flammable gas(es)?

– What is the limit ratio of a defined mixture of an inert gas (or a defined mixture of several inert gases) and a flammable gas (or a defined mixture of several flammable gases) for which the resulting mixture can reach its explosion limit and thus form an ATEX when mixed with air?

All mixtures considered here combine three types of gases:

– a flammable gas;

– an oxidizing gas, most often air (in reality, it is the oxygen contained in air that constitutes the oxidizing gas);

– and an inert gas.

Answers to these questions can generally be obtained, in a simple but reliable way, using a triangular representation of the ternary system that combines these three types of gases.

To characterize the flammability in air of a mixture of one or more flammable gases in the presence of an inert gas, it must be first considered successively:

– the determination of the LEL and UEL in air of each flammable gas alone;

– the determination of the LEL and UEL in air of a mixture of two or more flammable gases;

– the determination of the LEL and UEL in air of one or more flammable gases in the presence of one or more inert gases.

1.1.1.7. *Determination of the LEL and UEL in air of a flammable gas alone*

LEL(A) and UEL(A) shall be the respective values of the LEL and UEL in air of a flammable gas (A).

There is no simple numerical relationship that can accurately determine LEL(A) and UEL(A) from the physicochemical characteristics of (A). These parameters are only accessible through experience.

LEL(A) is obtained by experimentally testing, under specified conditions, the flammability of different mixtures of (A) and air and by framing the couple of mixtures of compositions as close as possible, such as:

– one is the flammable mixture with the lowest (A) content;

– the other is the non-flammable mixture with a content of (A) as close as possible to that of the previous flammable mixture;

– in European standards, LEL(A) is then considered to be the (A) content of the previous non-flammable mixture.

Similarly, UEL(A) is obtained by determining the couple of mixtures having their compositions as close as possible and such as:

– one is the flammable mixture with the highest (A) content;

– the other is the non-flammable mixture with an (A) content as close as possible to that of the previous flammable mixture.

UEL(A) is then considered to be the (A) content of the previous non-flammable mixture.

Note that U.S. standards consider the limit contents of flammable gas for which the mixtures are flammable as the value of the LEL or UEL.

Note that the LEL and UEL are dependent on temperature and pressure conditions: generally, when temperature and/or pressure increases, the LEL decreases and the UEL increases (the influence of temperature on LEL and UEL has already been discussed in section 1.1.1.5.2).

For example, if (A) = CO:

– the commonly accepted values of LEL(CO) and UEL(CO) are, respectively, 13.7% volume and 70.2% volume;

– the flammability range of CO–air is therefore the domain of CO content in CO–air mixtures between 3.7 and 70.2% volume.

1.1.1.8. *Determination of LEL and UEL in air of a mixture of two or more flammable gases*

It is common to have to know the values of the LEL and UEL of mixtures containing several fuels: the empirical law of Le Châtelier (INRS ED911 2004) enables us to determine approximate values of these limits, based on the composition of the mixture and the values of the LEL and UEL of each constituent taken separately.

Let us consider:

– a mixture A of several fuels A1, A2, A3, etc. and the known values of LEL(A1), LEL(A2), LEL(A3), respectively;

– (a1), (a2), (a3) … as the contents of A1, A2, A3, in mixture A1–A2–A3…

The values (a1), (a2), (a3)… are bound by relation [1.5]:

$$(a1) + (a2) + (a3) + \ldots = 100; \qquad [1.5]$$

– LEL(A), the LEL in air of the mixture A1–A2–A3.

The value of LEL(A) is obtained from the following relationship [1.6] (all quantities are expressed in percentage volume):

$$\text{LEL(A)} = 100/((a1)/\text{LEL(A1)} + (a2)/\text{LEL(A2)} + (a3)/\text{LEL(A3)} + \ldots) \qquad [1.6]$$

Similarly, the UEL in air of the A1–A2–A3 … mixture, UEL(A), is obtained from (a1), (a2), (a3)…, and from the UEL of each constituent, depending on the relationship [1.7]:

$$\text{UEL(A)} = 100/((a1)/\text{UEL(A1)} + (a2)/\text{UEL(A2)} + (a3)/\text{UEL(A3)} + \ldots) \qquad [1.7]$$

For example, for a defined mixture CO–CH_4, the contents of which are, respectively, (CO) = 18 and (CH_4) = 82% volume, the implementation of Le Châtelier's law needs the LEL and UEL of CO and CH_4:

– LEL and UEL of CO are 13.7% volume and 70.2% volume, respectively;

– LEL and UEL of CH_4 are 4.3% volume and 16.3% volume, respectively.

The calculation of LEL and UEL of this mixture gives:

LEL(CO-CH_4) = 100/(18/13.7 + 82/4.3) = 100/(1.3 + 19.1) = 100/20.4 = 4.9% volume

LSE(CO-CH_4) = 100/(18/70.2 + 82/16.3) = 100/(0.26 + 5.0) = 100/5.26 = 19% volume

Finally, the flammability range of this CO–CH_4 mixture in air is between 4.9% and 19% volume.

1.1.1.9. *Determination of LEL and UEL in air of a flammable gas in the presence of an inert gas*

1.1.1.9.1. Introduction to ternary diagrams

Any mixture consisting of a flammable gas (A), air (B) and inert gas (C) shall belong to a ternary system, which may be represented by an equilateral triangle with the summits A, B and C (see Figure 1.2):

– each point A, B and C represents the gas (A) alone, air alone and the gas (C) alone, respectively;

– each side of the triangle represents a binary mixture of the two components corresponding to each end of this side. For example, the AB side represents all air and gas (A) mixtures;

– each D point taken within the ABC triangle represents a ternary mixture of (A), air and (C);

– the content (a) of (A) in the mixture (D), expressed in percentage volume, is determined by the ratio of the lengths of segments D'B and AB, where D' is the intersection point of the parallel to the BC side passing through D (relationship [1.8]):

$$a = 100\,(D'B)/(AB) \qquad [1.8]$$

– similarly, the content (b) of the mixture (D) in air (the content (c) of the mixture (D) in (C), respectively) is determined by the ratio of the lengths of segments D''C and BC (D''A and CA, respectively), where D'' (D''', respectively) is the intersection of the parallel to the AC side (AB, respectively) passing through D, (relationships [1.9] and [1.10], respectively):

$$(b) = 100\ (D''C)/(BC) \qquad [1.9]$$

$$(c) = 100\ (D''A)/(AC) \qquad [1.10]$$

– finally, (a), (b) and (c) are bound by relationship [1.11]:

$$(a) + (b) + (c) = 100 \qquad [1.11]$$

1.1.1.9.2. Examples of ternary flammability diagram

Case of the CO–air–N_2 system

As a first example, Figure 1.2 shows the triangular flammability diagram of the CO–air–N_2 ternary system:

– A, B and C summits represent CO alone, air alone and N_2 alone, respectively;

– D is a representative point of any mixture taken within the triangle, the points D', D'' and D''' corresponding respectively to the intersections with the AB, BC and AC sides of the parallels with the BC, AC and AB sides passing through D.

Based on the respective contents of the mixture represented by the point D in CO, i.e. (a) = 39% volume, in air, i.e. (b) = 10% volume, and in N_2, i.e. (c) = 51% volume, equality (12) is verified:

$$(a) + (b) + (c) = 100 \qquad [1.12]$$

On the side AB of the triangle ABC, points E and F correspond respectively to the LEL and UEL of (A) in air.

The point G between E and F is representative of a flammable stoichiometric mixture (A)–air. It means that the contents of (A) and air are such that there is no excess of air or (A) and the combustion of (A) is complete.

In the case of CO–air–N_2 mixtures, the stoichiometric (or oxy-balanced) combustion reaction in air, in the absence of added N_2, is written as follows (relationship [1.13]):

$$0.295\ CO + 0.705\ air\ (0.148\ O_2 + 0.557\ N_2) = 0.295\ CO_2 + 0.557\ N_2 \qquad [1.13]$$

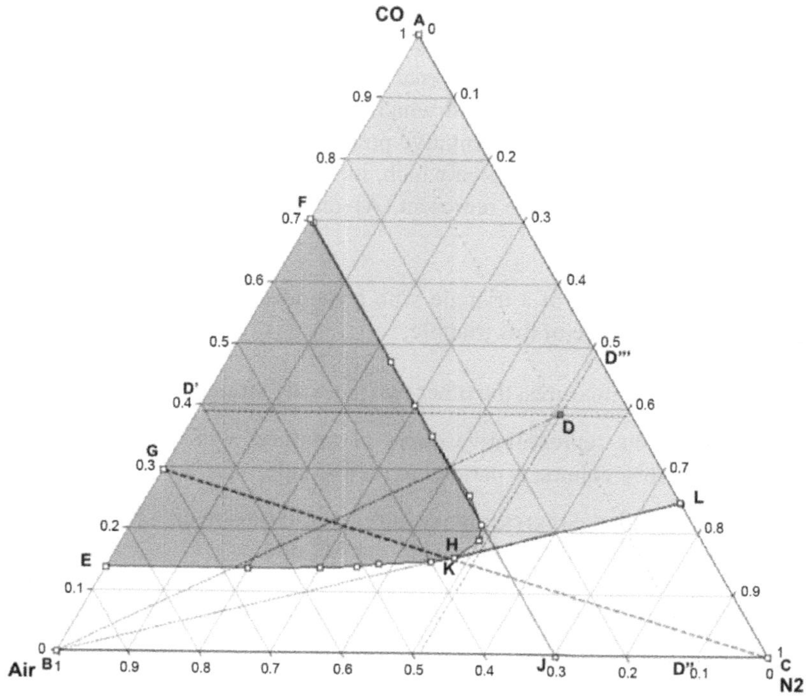

Figure 1.2. *Triangular flammability diagram of the ternary CO–air–N_2 system, at 20°C, 1 bar*

The CO content at point G, which corresponds to the stoichiometric composition, is therefore (a) = 29.5% volume.

The points of the segment GC are representative of all stoichiometric composition mixtures, whatever their content in inert (C).

For a given volume V(G) of the mixture (A)–air, represented by the point G and for a given volume V(C) of the inert gas alone, represented by point C, the representative point of a mixture of the volumes V(G) and V(C) belongs to the segment GC, whatever the values of V(G) and V(C).

Since point G represents an explosive stoichiometric mixture, while point C represents the inert alone, there is necessarily a point H of the segment GC that represents a mixture at the explosion limit: it corresponds to the explosive stoichiometric mixture with the highest inert content (C).

For the CO–air–N_2 mixtures, the CO content at point H equals 15.6% volume.

The existence of points E, F and H enables us to assert the existence of a curve which passes through these points and which separates all representative points of flammable mixtures from all representative points of non-flammable mixtures. This curve delineates the flammability area which, in Figure 1.2, corresponds to the dark gray area. All points on the curve represent (A)–air–(C) mixtures that are just at the explosion limit.

This area has the shape of a triangle with a rounded summit closest to point C. The segment EF is one side of this triangle.

For the CO–air–N_2 mixtures, the other two sides are straight:

– on one hand, the representative curve of mixtures with a proportion of CO between 13.7 and 16% volume is parallel to the BC side;

– on the other hand, the representative curve of mixtures with a proportion of CO between 21 and 70.2% volume is parallel to the AC side.

1.1.1.10. *Flammability characteristics of flammable ternary mixtures*

1.1.1.10.1. Oxygen limit content

For any given flammable gas (A) and for a given inert gas (C), there is an air limit content (LairC) below which no mixture (A)–air–(C) is flammable, regardless of its (A) content. This air limit content corresponds to the air content of the point J (J is the intersection of the BC side with the parallel to the AC side passing through F). No (A)–air–(C) mixture is flammable if its representative point is located between this parallel and the AC side. This air limit content (LairC) is therefore equal to the ratio CJ/CB of the lengths of the segments CJ and CB.

The oxygen limit content (LOC) is derived from the relationship [1.14]:

$$LOC = 0.2095 \times LairC \qquad [1.14]$$

where 0.2095 is the value of the oxygen content of air.

It should be noted that knowledge of the LOC value is essential to the prevention of explosions through the implementation of the inerting process, which consists of keeping the oxygen content of any mixture containing a flammable gas (A) permanently below the LOC.

In the case of the ternary CO–air–N_2 system, (LairC) = 30% and (LOC) = 6.2% volume.

1.1.1.10.2. Safety limit ratio of flammable and inert gas contents

Figure 1.2 shows that a mixture can only be flammable if its flammable gas and inert gas contents are sufficiently high and low, respectively.

The maximum value of the ratio (c)/(a) that characterizes the explosion limit of mixtures (A)–air–(C) is equal to the ratio LC/LA of the lengths of the segments LC and LA, where L is the intersection point of the straight line passing through B, which is tangent in K to the flammability curve.

According to Figure 1.2, the length of the LC segment is 24.5% volume and the safety limit ratio of the H_2 and CO_2 contents (c)/(a) = 3.1.

In addition, the KL segment shows two zones in the ABC triangle:

– The area of points above the BL but not within the flammability area: they are therefore representative of non-explosive mixtures (point D in Figure 1.3 is one of the points in this area). However, if air is added to the mixture represented by point D, the representative point of the resulting mixture describes the BD segment that cuts the flammability area: therefore, there is a range of air dilution ratios of mixture D for which the resulting mixture becomes flammable, and this area has been colored in light gray.

– The white area that is bounded by:

- the flammability area between points E and K;
- the segment KL;
- the segment LC;
- the side BC of the triangle;
- the segment BE.

All points within this white area are outside the flammability area and are therefore representative of non-flammable mixtures. Furthermore, there is no air dilution ratio of these mixtures that could lead to the formation of a flammable mixture.

1.1.1.10.3. Case of the H_2–air–N_2 system

As a second example, Figure 1.3 shows the explosion diagram of the ternary H_2–air–N_2 system, based on data from Molname et al. (2003).

In the case of H_2–air–N_2 mixtures, the stoichiometric (or oxy-balanced) combustion reaction in air, in the absence of added N_2, is:

$$0.295\ H_2 + 0.705\ \text{air (or } 0.148\ O_2 + 0.557\ N_2) = 0.295\ H_2O + 0.557\ N_2 \quad [1.15]$$

The H_2 content at point G, which corresponds to the stoichiometric composition, is therefore (a) = 29.5% volume. Moreover, the H_2 content at H is 8.4% volume and the safety limit ratio of H_2 and N_2 contents is 15.7.

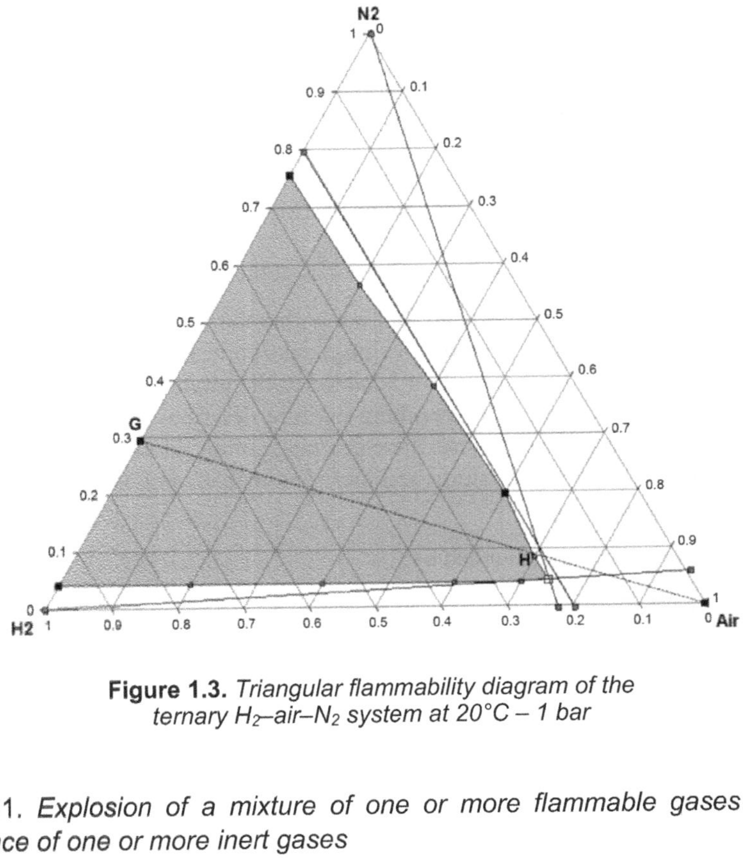

Figure 1.3. *Triangular flammability diagram of the ternary H_2–air–N_2 system at 20°C – 1 bar*

1.1.1.11. *Explosion of a mixture of one or more flammable gases in the presence of one or more inert gases*

In some cases, the flammability of a mixture used during the operation of a process in which a hydrocarbon fuel is burned may be estimated for conditions of under-oxygenation.

For example, such a situation can be found in incinerators, where combustion may be poorly regulated. In this case, different gases are produced, some of which are flammable (H_2, CO, CH_4 for the main ones) and others are inert (CO_2 and H_2O for the main ones), so that the resulting mixture contains all these gases, plus N_2, or even a few percent of residual O_2. To estimate the flammability of this mixture, it would be necessary to know the flammability in air of a defined mixture of flammable gases H_2, CH_4 and CO, in the presence of a defined mixture of inert gases

CO_2, H_2O and N_2. However, no flammability diagram was experimentally determined for a defined ternary system combining all these gases.

The US Bureau of Mines has published a method (USBM B503 1953) to characterize the flammability of a mixture consisting of all or part of the above gases, as well as the possibility of obtaining a flammable mixture by an air dilution of the initial mixture.

This question arose first in the coal mining operation, since it is the same gases that are produced when a coal mass starts a burning fire in the confined environment of a gallery and it is important to know how to put out the fire taking into account the composition of the mixture of the gases emitted, while avoiding their dilution by air which could lead to an explosion.

An application of this method is described further, in the case studies of explosions in cupola facilities (see Part 2 case 11).

1.1.2. *Case of an ATEX consisting of a combustible dust dispersed in air*

A large number of combustible dusts, including virtually all those used in the agro-food industry, can lead to dust explosions. In France, explosions in grain elevators in Metz in 1982 and Blaye in 1997 marked memories.

As long as the chemical composition of a combustible solid or mixture of solids is known, it is possible to write its combustion equation in the same way as that of a gas or vapor emitted by a flammable liquid and therefore to determine the mass of that solid which must be dispersed as a dust in 1 m^3 of air to give rise to a stoichiometric combustion.

The flammability in air of a combustible dust is characterized by the same parameters as for a flammable gas, but with the main following differences:

– the concepts of LEL and UEL exist, but, unlike those of gases and vapors, these characteristics depend on some other characteristics of the dust;

– its granulometry;

– its moisture;

– its ignition energy.

The UEL corresponds to a mixture of several kg/m^3 of dust dispersed in air, and there is no real interest to know it precisely since, for such high contents, the combustion is not total and leaves a large fraction of partially unburned dust: it is as if the flame propagated through mixtures having contents higher than UEL which then should no longer be considered.

On the contrary, LEL is generally in the order of a few tens of g/m^3 and is dependent on the ignition energy of the ATEX. It can be considered as the maximum permissible level of dusting, inside facilities where layers of dust can be deposited.

Actually, it must also be considered that the dust which participates to the combustion may not be initially and wholly dispersed, in so far as it is the incipient combustion of the dispersed fraction and the resultant blowing which favor the dispersion of the not yet dispersed fraction. This is the reason why the flame of a dust explosion can propagate over very large distances according to the following mechanism, when the ATEX is formed progressively in a very elongated shape (pipe, mine gallery, grain storage silo, etc.):

– a sufficient volume of ATEX consisting of a dust dispersed in air is formed locally and is ignited;

– the combustion generates a blowing that disperses in air the dust that was initially as a layer, near the initial ATEX;

– this dispersed dust forms an ATEX further than the initial one, which the flame reaches and through which it propagates;

– the dispersion and combustion of the dust initially as a layer repeats itself from close to near, until the amount of the dispersed dust becomes too small;

– this phenomenon has been particularly studied in the case of coal mines, where it is often a gaseous ATEX made of firedamp mixed with air that gives rise to initial combustion.

According to this mechanism, a dust explosion can cause destructive effects at distances far from the origin of explosion, which is thus all the more difficult to characterize.

The minimum ignition energy (MIE) of a dust is highly dependent on its granulometry and moisture content and is worth a few millijoules to a few joules, which is significantly higher than for a flammable gas or vapor (see section 1.1.1.2). Depending on the value of MIE, the ignition may or may not result from the energy released by electrostatic phenomena. Finally, a solid dust can result in a self-ignition, according to the following mechanism, when it is in the state of a layer deposited on a sufficiently hot surface:

– a layer of dust is a juxtaposition of grains resting on each other and whose interstices are filled with air;

– the initial temperature of the layer is assumed to be sufficient to initiate, at the surface of the grains, an oxidation, which releases heat;

– since the layer is a low heat-conductive medium, the transmission of this heat by conduction from the inside of the layer to the outside is slow enough to allow an increase in the temperature of the grain;

– this heating accelerates the oxidation of the grain and the phenomenon can trigger a self-ignition of the layer, in the form of a combustion process, with the appearance of incandescent zones where the temperature reaches 500 to 1,000°C.

Accidents in which a dust explosion is initiated by a phenomenon of self-ignition of a layer are frequent (two such cases are dealt with in Part 2 case 16).

1.1.3. *Case of a hybrid ATEX*

A hybrid ATEX consists of a mixture containing at least one flammable gas or vapor emitted from a flammable liquid and a solid combustible dust.

For some accidents, it has been possible to demonstrate that the mechanism had definitely involved the explosion of a hybrid ATEX. Indeed, such a mechanism is entirely possible and, for example, it could be invoked when a dust, impregnated with a flammable solvent, has to be dried in a hot airflow. The explosive properties of the dust must take into account the coexistence of the dust and the solvent of impregnation.

In particular, the LEL depends on the respective LEL's of the dust and the solvent, with an additivity law that could be similar to the Le Châtelier law. As for the value of MIE, the LEL should be considered to be close to that of the pure solvent, since when the explosion is initiated, it is preferably the vapor it emits that is present in ATEX.

1.1.4. *Evaluation of the released energy E_{lib}*

During the explosion of an ATEX containing a mass of combustible m, whose combustion enthalpy is ΔH_{comb}, the released energy E_{lib} is obtained from the relationship [1.16]:

$$E_{lib} = m\Delta H_{comb} \qquad [1.16]$$

where E_{lib} is the energy released during the explosion (in J), m is the combustible mass (in kg) and ΔH_{comb} is the enthalpy of the combustion (in J/kg).

The value of ΔH_{comb} depends on the composition of the combustible–oxidizer mixture, i.e. the proportions of air and combustible in this mixture. In fact, if the composition is not stoichiometric, it is because one of the reagents is in excess, so that, like the products of the reaction, this reagent is heated to the flame temperature and consumes energy without producing it.

Next, the case of a stoichiometric air–butane mixture, LEL and UEL mixtures are successively examined.

1.1.4.1. *Case of a stoichiometric air–butane mixture*

1.1.4.1.1. Composition of the mixture of reagents and products

Let us consider an ATEX consisting of a stoichiometric air–butane mixture.

Among the combustion products of this ATEX, H_2O and CO_2 are the most abundant and the complete combustion reaction of one mole of butane can be written as follows:

$$C_4H_{10} + 6.5\ O_2 + 24.5\ N_2 = 4\ CO_2 + 5\ H_2O + 24.5\ N_2 \quad [1.17]$$

But it is established that other chemical species are also present in the flame: derived from dissociation reactions of the main products, which are thermochemical equilibriums, some of them are unstable and it would be difficult to isolate them, their content in the mixture of the combustion products is very low (except CO, which can be abundant, especially when combustion occurs in more under-oxygenated environments). More generally, combustion in highly under-oxygenated environments of hydrocarbon molecules produces solid C and that of molecules containing halogen X atoms (with X = F, Cl, Br or I), produces HX compounds, or even COX_2, when the medium is highly oxygenated. It should also be added that the above reaction is simplified because, in the mixture of reagents, the presence of the following two inert gases, argon and CO_2, is not taken into account. However, their content in the mixture is very low (about 1% volume), so this presence has no significant effect on the thermochemical properties of the mixture.

If these different chemical species are omitted, an ATEX containing 1 mole of C_4H_{10} for 6.5 moles of oxygen and 24.5 moles of nitrogen (the equivalent of 31 "moles of air") contains 32 moles of gas and corresponds to an "oxy-balanced" or "stoichiometric" mixture.

Expressed in percentage volume (or in percentage molar), the content of butane C_{st} in a stoichiometric mixture is worth (relationship [1.18]):

$$C_{st} = 100/(1 + 6.5 + 24.5) = 100/32 = 3.125\% \text{ volume.} \quad [1.18]$$

Written for 1 mole of stoichiometric air–butane mixture, the complete combustion equation becomes (see equation [1.19]):

$$0.031\ C_4H_{10} + 0.203\ O_2 + 0.766\ N_2 = 0.124\ CO_2 + 0.155\ H_2O + 0.766\ N_2 \quad [1.19]$$

1 mole 1.045 mole

Finally, the contents of each reagent or product expressed in percentage volume in the reagent and product mixture are as follows:

– Reagents:
  - C_4H_{10} 0.031% volume;
  - O_2 0.203% volume;
  - N_2 0.766% volume.
– Products:
  - CO_2 11.9% volume;
  - H_2O 14.8% volume;
  - N_2 73.3%.

1.1.4.1.2. Calculation of combustion enthalpy ΔH_{comb}

The combustion energy ΔH_{comb} is calculated by the following relationship [1.20]:

$$\Delta H_{comb} = \Sigma(n_{pi}(\Delta H_{f0})_{pi}) - \Sigma(n_{ri}(\Delta H_{f0})_{ri}) \quad [1.20]$$

where:

– n_{pi} is the number of moles of the product (i);
– n_{ri} is the number of moles of the reagent (i);
– $(\Delta H_{f0})_{pi}$ is the standard enthalpy of formation of the product (i);
– $(\Delta H_{f0})_{ri}$ is the standard enthalpy of formation of the reagent (i).

In the hypothesis of a simplified writing, the only products are H_2O and CO_2 (for pure bodies O_2 and N_2, ΔH_{f0} is zero, by convention), whose values of (ΔH_{f0}) are provided in the database (Component Plus Database of ProSim 2018):

– $\Delta H_{f0}(C_4H_{10})$ = +126 kJ/mole;

– $\Delta H_{f0}(CO_2)$ = -393 kJ/mole.

For water, ΔH_{f0} depends on its physical state, liquid or gaseous (steam):

– $\Delta H_{f0}(H_2Ogas)$ = -242 kJ/mole;

– $\Delta H_{f0}(H_2Oliq)$ = -285 kJ/mole.

For a stoichiometric mixture, the combustion enthalpy ΔH_{combst} is calculated from equation [1.20], where the numbers of moles of the products and reagents are those that appear in equation [1.17] and where water is taken in liquid state:

$$\Delta H_{combst} = 4 \times (-393) + 5\,(-285) - (126) = -2.9 \text{ MJ/(mole of butane)}$$

$$\Delta H_{combst} = -2.9/0.058 = -50.1 \text{ MJ/(kg butane)}$$

It is remarkable that this value represents a little more than 10 times the energy released by the decomposition of TNT, i.e. 4.69 MJ/kg.

1.1.4.2. *Consideration of dissociation equilibriums*

In addition to the main products of butane combustion, namely CO_2 and H_2O, the chemical species identified in the mixture of products are, from the most stable to the most unstable, CO, H_2, O_2, NO, OH, H and O.

The ExplPress combustion model, which was developed as part of the European project (for the SAFEKINEX EU project, see ExplPress (2006)), provides the values of the contents of each chemical species in the mixture of the combustion products (percentage volume), which make it possible to write this reaction for 1 mole of mixture of reagents, which are given in Table 1.3.

A comparison of the above H_2O and CO_2 contents with those corresponding to the simplified writing of the combustion reaction (see section 1.1.4.1.1) shows that the former are only slightly less than the latter:

– H_2O content decreases from 14.8% volume to 13.6% volume;

– the CO_2 content decreases from 11.9% volume to 9.4% volume; the difference is greater in this case, and it is due to the fact that, in the writing, CO production is neglected, while it represents 2.2% volume of the products mixture.

Products	Content
CO_2	9.4
H_2O	13.6
N_2	72.0
CO	2.2
H_2	0.49
O_2	0.9
NO	0.57
OH	0.66
H	0.09
O	0.09

Table 1.3. *Nature and content of each chemical species, contained in the products mixture, resulting from the combustion of a stoichiometric air–butane mixture, when the dissociation equilibriums of certain species are taken into account*

1.1.4.3.1. Cases of mixtures corresponding to LEL and UEL of butane in air

The LEL and UEL of butane are worth 1.8% and 8.4% volume, respectively (INRS ED 911 2004).

Case of a LEL air–butane mixture

The simplified writing of the combustion equation of an LEL air–butane mixture is as follows (see equation [1.21]):

$$C_4H_{10} + 11.43\ O_2 + 43.13\ N_2 = 4\ CO_2 + 5\ H_2O + 4.93\ O_2 + 43.13\ N_2 \qquad [1.21]$$

An LEL air–butane mixture contains 54.56 "moles of air" for 1 mole of butane and therefore contains a total of 55.56 moles of gas.

The combustion enthalpy $\Delta H_{CombLEL}$ is calculated from equation [1.17] where the numbers of moles of the reagents and combustion products are those of equation [1.21].

On one hand, if $\Delta H_{CombLEL}$ is expressed in reference to the butane alone, it is worth the same value as that of a stoichiometric mixture:

$$\Delta H_{CombLEL} = -2.9/0.058 = -50.1 \text{ MJ/(kg butane)}.$$

On the other hand, it can be expressed in reference to an LEL air–butane mixture, i.e.:

$\Delta H_{CombLEL}$ = - 2.9/55.56= -0.052 MJ/(mole of an LEL air–butane mixture).

Thus, the combustion heat, $\Delta H_{CombLEL}$, of an air–butane LEL mixture, is worth about half the combustion heat ΔH_{combst} mixture, because in an LEL mixture, there is about half the amount of butane compared with a stoichiometric mixture.

Case of a UEL air–butane mixture

The simplified writing of the combustion equation of a UEL air–butane mixture is as follows (see equation [1.22]) if it is supposed that the excess of butane remains unburned and unchanged):

$$2.84\ C_4H_{10} + 6.5\ O_2 + 24.5\ N_2 = 1.84\ C_4H_{10} + 4\ CO_2 + 5\ H_2O + 24.5\ N_2 \qquad [1.22]$$

An UEL air–butane mixture contains 31 "moles of air" for 2.84 moles of butane, and therefore contains a total of 33.84 moles of gas.

The combustion enthalpy $\Delta H_{combUEL}$ is calculated from equation [1.17] where the number of moles of the reagents and combustion products are those of equation [1.22]:

Again, when $\Delta H_{combUEL}$ is expressed in reference to the butane alone, it is worth the same value as that of a stoichiometric mixture, i.e.:

$\Delta H_{combUEL}$ = -2.9/0.058 = -50.1 MJ/(kg butane).

But, it can be expressed in reference to an air–butane UEL mixture, i.e.:

$\Delta H_{combUEL}$ = -2.9 x 2.84/33.84 = -0.24 MJ/(mole of a UEL air–butane mixture).

Calculation of the flame temperature T_f of an ATEX explosion

To carry out the investigation of an explosion involving an ATEX present in a containment, it is often necessary to estimate the maximum pressure that was or could have been reached in that containment.

It is then necessary to estimate several characteristics of the explosive system, first of all the flame temperature T_f.

Definition and characteristics of a flame temperature T_f

The flame temperature T_f is defined as the "explosion adiabatic temperature", i.e. the maximum temperature that can be reached by the products of the combustion reaction:

– in fact, it is a fictitious temperature since it is the one that would have been reached if all the energy released would have been used to heat the products of the reaction, without loss towards the outside, i.e. in particular without any heat exchange between the mixture of the hot combustion products and the wall of the containment;

– it is when an ATEX is ignited in the center of a spherical containment that this condition is best met and the actual flame temperature is closest to T_f;

– when the flame remains away from the containment wall throughout the combustion process and only approaches it at the end, there is no transfer of energy to the wall by conduction or convection, but only by radiation;

– the hypothesis is then fairly well verified since the measured explosion pressure of an ATEX having a given composition is only slightly lower than its calculated value.

Determination of T_f from the simplified combustion equation

The T_f calculation is based on a thermal balance in which the energy used to heat the products of the reaction, from the initial temperature T_0 to $T_f = T_0 + \Delta T$, is just worth the enthalpy ΔH_{comb} of the simplified combustion reaction [1.23].

$$T_f = T_0 + \Delta T = T_0 + \Delta H_{comb}/(\Sigma(np)_i(cp)_i) \qquad [1.23]$$

where:

– T_0 is the initial ATEX temperature;

– ΔH_{comb} is the enthalpy of the combustion reaction, calculated with equation [1.20];

– $(cp)_i$ is the thermal molar capacity of the product (i) in the combustion reaction (J/mole.K);

– $(np)_i$ is the number of moles of the product (i) in the combustion reaction.

The values of $(cp)_i$ of the most abundant species in the combustion products are given as a function of T in the database (Component Plus Database of ProSim 2018).

Determination of the explosion pressure P_{ex} of an ATEX

From the T_f value of an ATEX having a given composition, and from the volume V of a containment of this ATEX, it is possible to determine its explosion pressure P_{ex}, i.e. the maximum pressure of the mixture of combustion products, which is reached in the containment at the end of the reaction.

The initial and final systems are supposed to be defined by the nature of the reagents and products, as well as by the composition of the mixture of reagents and of the mixture of products, according to the relationships between the following parameters:

– nr_i is the number of moles of the reagent (i);

– $Nr = \Sigma(nr_i)$ is the total number of moles of the reagents;

– Mr_i is the molar mass of the reagent (i) (in g/mole);

– Meq_r is the equivalent molar mass of the mixture of reagents (in g/mole);

– np_i is the number of moles of the combustion product (i);

– $N_p = \Sigma(np_i)$ is the total number of moles of the of the combustion products (in g/mole);

– Mp_i is the molar mass of the reaction product (i) (in g/mole);

– Meq_p is the equivalent molar mass of the mixture of combustion products (in g/mole).

The initial and final systems are supposed to follow the law of perfect gases, as long as the pressure of the mixture of reagents or gaseous combustion products does not exceed 100 bar.

The characteristics of the initial and final systems are then linked by the following relationships [1.24] and [1.25]:

$$P_0 = NrRT_0/V \quad [1.24]$$

where:

– P_0 is the initial pressure (in Pa);

– R is the gas constant = 8.31 J/mole.K;

– T_0 is the initial system temperature (in K);

– V is the volume of the containment (in m^3);

and

$$P_{ex} = NpRT_f/V \qquad [1.25]$$

where:

– P_{ex} is the explosion pressure (in Pa);

– T_f is the flame temperature (in K).

The ratio $\alpha = Np/Nr$ is called the "expansion ratio"; it is the ratio of the total number of moles of the combustion products to the total number of moles of the reagents. For hydrocarbon fuels, this ratio is close to 1.

The equivalent molar mass Meq_r of the mixture of reagents is obtained from the molar mass Mr_i and the number of moles nr_i of each reagent (i) in the mixture of reagents (relationship [1.26]):

$$Meq_r = \Sigma(nr_iMr_i)/\Sigma(nr_i) \qquad [1.26]$$

Similarly, the equivalent molar mass Meq_p of the mixture of combustion products is obtained from the molar mass Mp_i and the number of moles of each combustion product (i) in the mixture of combustion products (relationship [1.27]):

$$Meq_p = \Sigma(np_iMp_i)/\Sigma(np_i) \qquad [1.27]$$

Finally, the mass Mr of the mixture of reagents (respectively, the mass Mp of the mixture of combustion products) is the product of its equivalent molar mass Meq_r (respectively Meq_p) by the number of moles Nr (respectively, Np) of this mixture (relationships [1.28] and [1.29]):

$$Mr = Meq_r \text{ x } Nr \qquad [1.28]$$

$$Mp = Meq_p \text{ x } Np \qquad [1.29]$$

The value of P_{ex} is calculated from the relationship [1.30], which is deduced from the relationships [1.28] and [1.29]:

$$P_{ex} = (NpT_f/NrT_0)P_0 = \alpha\, P_0\, (T_f/T_0) \qquad [1.30]$$

Different characteristics of air–butane mixtures, taken at 20°C, 1 bar and having butane contents between LEL and UEL, have been calculated by using the ExplPress (2006) model. For butane contents c (in percentage volume) corresponding to LEL, stoichiometry and UEL, the following values are given in Table 1.4:

– expansion ratio α;

– combustion enthalpy ΔH_{comb} (in MJ/kg butane);

– T_f flame temperature (in K and °C);

– P_{ex} explosion pressure (in bar).

Type of air–butane mixture	Butane content c (% volume)	Expansion ratio α	ΔH_{comb} (in MJ/kg butane)	T_f		P_{ex}
				(K)	(°C)	(bar)
LEL mixture	1.8	1.03	50.1	1,996	1,723	7.0
Stoichiometric mixture	3.1	1.05		2,630	2,357	9.55
UEL mixture	8.4	1.04		1,449	1,176	7.3

Table 1.4. *Values of the expansion ratio α, the combustion heat ΔH_{comb}, the flame temperature T_f and the explosion pressure P_{ex} of air–butane mixtures at different contents c in butane*

Figure 1.4 shows the variation curves of the P_{ex} explosion pressure (black curve) and of the flame temperature T_f (red curve) as a function of the content c of butane in air–butane mixtures at 20°C below 1 bar, for a composition between LEL and UEL.

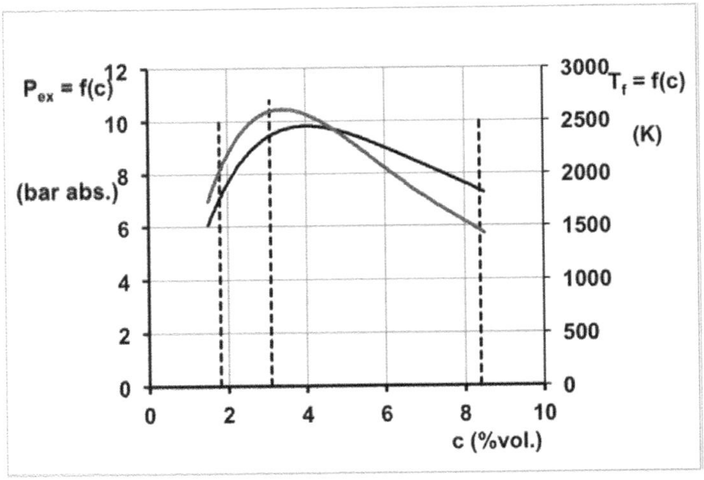

Figure 1.4. *Variation curves of the explosion pressure P_{ex} (black curve) and of the flame temperature T_f (red curve), as a function of the butane content c of an air–butane mixture at 20°C below 1 bar. For a color version of this figure, see http://www.iste.co.uk/chaineaux/mechanisms.zip*

1.1.4.3.2. Case of an experimental explosion of a 10% volume methane-air ATEX

Experimental explosions of a 10% methane-air ATEX

The simplified equation [1.31] of the combustion reaction of methane in air is written:

$$CH_4 + 2\ O_2 + 7.55\ N_2 = CO_2 + 2\ H_2O + 7.55\ N_2 \quad [1.31]$$

A methane content of 10% volume is slightly higher than the stoichiometric content, i.e. 9.5% volume.

Figure 1.5 shows the variation with time of the internal overpressure P_{ex} in a 2 m^3 sphere (78 cm radius), containing an air–methane mixture of 10% methane, prepared at 20°C, 1 bar and ignited in the center of the sphere.

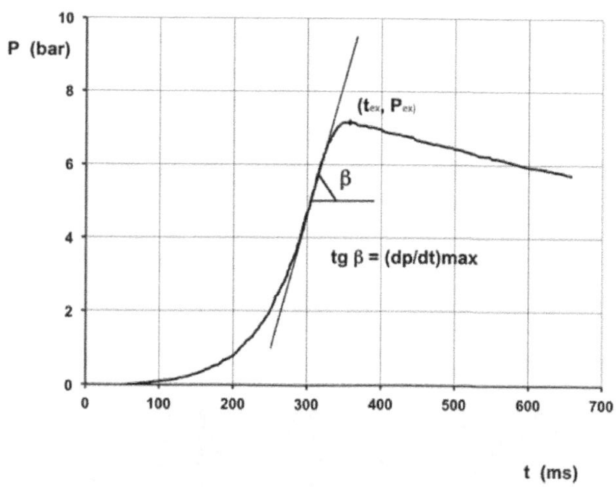

Figure 1.5. *Variation with time of the internal overpressure P_{ex} in a 2 m^3 sphere (78 cm radius), within which an air–methane mixture of 10% methane, prepared at 20°C, 1 bar, is ignited in the center of the sphere*

From Figure 1.5, the following characteristics can be determined:

– the measured relative explosion overpressure P_{ex} = 7.2 bar; this value corresponds to an absolute pressure P_{exabs} = 7.2 + 1 = 8.2 bar, which can be compared to the value calculated with the model, i.e. 9.05 bar;

– the time t_e taken to reach P_{ex}, i.e. 360 ms;

– the value of the maximum pressure rise $(dp/dt)_{ex}$: it is equal to tg β, where β is the angle of the tangent at the inflection point of the curve with the horizontal, i.e. approximately 10,000 bar/s.

The expansion ratio corresponding to the combustion of CH_4 is α = 1, and the relationship [1.30] becomes [1.32]:

$$T_f = T_0 \, (P_{ex}/P_0) \qquad [1.32]$$

For $T_0 = 293$ K and $P_{ex}/P_0 = 9.05$, $T_f = 2{,}652$ K.

1.1.4.3.3. Case of an experimental explosion of a stoichiometric air-propane ATEX

Stoichiometric propane–air explosion test

Description of the experimental setup

An ATEX consisting of a stoichiometric air–propane mixture, i.e. 4% propane in the mixture, is prepared in a cubical 50 m^3 containment (average edge length: 3.7 m). The six faces of the cube are made up of reinforced concrete for lower (ground), rear, and right and left sides and polyethylene (PE) sheets for front and top. The ATEX was ignited by an "electric match" on the vertical axis, at approximately 1.5 m from the ground. Flame propagation is filmed through the front of the cube at a rate of 24 fps.

Figure 1.6 shows five views of the flame that were drawn from the film for increasing times during the explosion and which are characteristic of a propagation of the flame in the regime of the deflagration.

t = 0.1 s

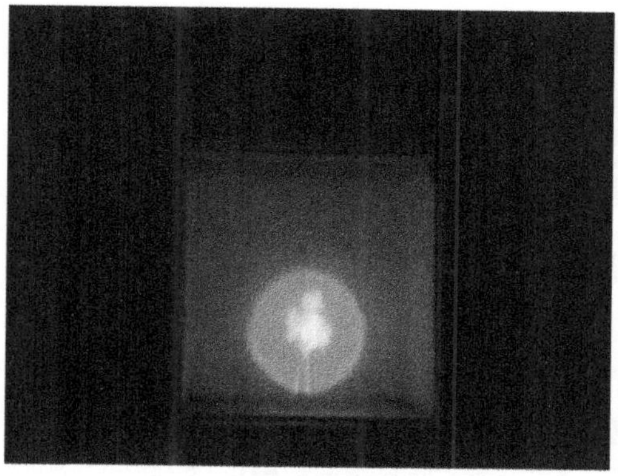

t = 0.3 s

t = 0.45 s

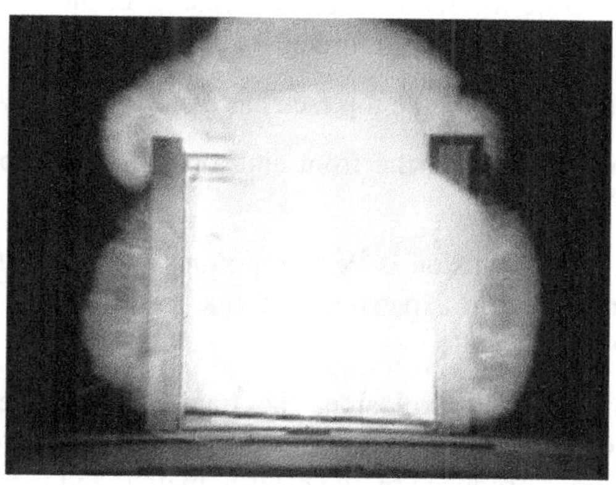

t = 0.65 s

t = 0.85 s

Figure 1.6. *5 views taken at increasing times during the explosion of a stoichiometric air–propane mixture, contained in a cubical containment of four rigid and two deformable faces. For a color version of this figure, see www.iste.co.uk/chaineaux/mechanisms.zip*

Interpretation

The following observations can be drawn from Figure 1.6:

– the flame is spherical, from its beginning to approximately 0.4 s, which means that it propagates at the same speed in all directions;

– at t = 0.45 s, its shape has become ovoid, as the flue gases (or burned gases), which are produced inside the flame, are less dense than the initial ATEX and are pushed by the buoyancy as an upward force;

– at this time, the PE sheet on the top face has already been severely deformed;

– a bit earlier than t = 0.65 s, the front and top faces are ruptured and the flue gases expand out of the cube;

– for t = 0.85 s, this expansion is very important and most of the initial ATEX has burned (at their maximum expansion, the flue gases occupy a roughly cubical volume of about 500 m^3);

– at the beginning of the explosion, the flame has a blue color because it corresponds to a stoichiometric ATEX (the orange color in the center of the cube corresponds to the fumes produced by the electric match) and then, for t greater than 0.3 s, the flame gets pink;

– at t = 0.65 s, the flue gases get in contact with the ambient air, and outside the cube, the flame is blue;

– finally, at t = 0.85 s, the mixture of the flue gases and air is very inhomogeneous and, its color combines blue and orange.

In addition, the presence of two faces of fragile material allows easy expansion of the flue gases so that, during the duration of the explosion, the pressure inside the flue gases remains practically equal to the one inside the initial ATEX, i.e. atmospheric.

If the test had been carried out in a cube with six rigid faces, the pressure would have increased under the following conditions:

– in the case of a fully resistant container, the explosion pressure P_{ex} would have reached a little less than 10 bar, the pressure rise profile having the aspect of that of Figure 1.6, with the same P_{ex} value, but with a longer time to reach it (about 1 s instead of 0.36 s);

– however, although it is made of reinforced concrete, the cube would have not been able of withstand such a pressure, which is equivalent to a force of about 100 daN exerted on each m^2 of wall;

– in reality, the walls of the cube would have been disintegrated and fragments of concrete would have been blown away, as soon as the pressure inside the cube would have reached a relatively low value, i.e. 1 bar.

1.2. Chemical systems other than ATEX

1.2.1. *Definition elements*

Systems other than ATEX can store energy in the chemical state and release that energy during a rapid chemical reaction:

– it may be a system consisting of two or more reagents likely to give rise to an exothermic chemical reaction;

– it may also be a single product which, under certain conditions, may give rise to an exothermic decomposition reaction:

- this is the case for certain solid products, for example, nitrated hydrocarbons such as trinitrotoluene (TNT), $C_7H_5N_3O_6$ or ammonium nitrate (NH_4NO_3),

- the characteristic of these products, which are commonly referred to as "explosives", is that their decomposition occurs in the absence of air (so it can also occur underwater),

- the products of the exothermic reaction are generally gases; these gases are heated to high temperatures (more than 1,000 K), and they undergo a thermal expansion which is at the origin of the effects of the explosion.

Certain substances that are gaseous at room temperature are also likely to decompose in the absence of air and are therefore similar to explosives, such as acetylene, ethylene oxide or hydrazine and its derivatives.

1.2.2. *Evaluation of E_{lib}*

The energy E_{lib} released during the explosion of chemical systems other than ATEX is:

$$E_{lib} = m\, \Delta H_{reac} \qquad [1.33]$$

where:

– E_{lib} is the energy released during the explosion (in J);

– ΔH_{reac} is the enthalpy of the reaction between several reagents or of the decomposition reaction of the single product; it is expressed in J/kg, i.e. in J/kg of system or in J/kg of one of the reagents, or in J/kg of the single product;

– according to the definition of ΔH_{reac}, m (in kg) is the mass of the system or the mass of one of the reagents or the mass of the single product.

E_{lib} also depends on the composition of the reagent mixture, for systems with two or more reagents. For example, for TNT, the value of ΔH_{reac} is close to 4.69 MJ/kg TNT.

1.2.3. *Flame propagation regimes in explosive system 1 or 2*

The explosion of a type 1 or type 2 explosive system is characterized by the production of a flame that can propagate in either of the following two regimes: the deflagration or the detonation. The characteristics of each regime are markedly different, both in terms of the circumstances in which they can be established and of the effects produced.

In the following, the different regimes are successively presented:

– the characteristics of each regime, in particular the differences in their mechanism and the aerial pressure wave they produce;

– the intermediate regime of the so-called "rapid" deflagration;

– the conditions for a possible passage from the deflagration to the detonation or to the "rapid deflagration".

1.2.3.1. *Characteristics of a detonation*

Whether the components of the explosive system are in solid state (e.g. TNT), liquid state or gaseous state, experience shows that a detonation can result from the decomposition or combustion of sufficiently reactive products: this reaction occurs in the detonation wave, which presents itself as an association, on a thickness of about 0.1 mm, of two distinct successive phenomena:

1) the downstream phenomenon (in the direction of propagation) is a shock wave which has the property of compressing very strongly the explosive system in which it propagates, and heating it sufficiently to initiate its reaction;

2) the upstream phenomenon is the flame where the reaction transforms the reagents into products, releasing the energy corresponding to the variation of the enthalpy of the mixture: it is this energy that keeps the shock wave going.

Inside the explosive system, the detonation wave propagates at a clearly supersonic constant speed (at least 1 km/s for gaseous systems and up to 10 km/s for solid systems). When the flame has consumed the entire explosive system, the reaction stops and the shock wave continues to propagate through the air, slowing down, until it becomes a sonic wave propagating at the sound speed in the air, i.e. 340 m/s.

1.2.3.2. *Characteristics of a deflagration*

In a deflagration of an ATEX (explosive system 1), the flame propagation results from the heating of the downstream environment (so-called "fresh gases") by the combustion products formed in the upstream environment (so-called "burned gases"), and this heating brings the activation energy necessary for the initiation of the fresh gases reaction. This mechanism repeats itself from close to near until the entire ATEX is consumed. At the beginning of its propagation, the flame speed does not exceed a few tens of cm/s and no shock wave is produced.

It should be noted that a deflagration is not restricted to gaseous systems and may also concern liquid or solid systems; in the case of a hunting ammunition, the desired decomposition regime of the cartridge is the deflagration and not the detonation.

1.2.3.3. *Transition from deflagration to detonation*

When a gaseous ATEX is contained in a very elongated containment, such as a pipe, and when this ATEX is ignited near the closed end of the pipe, the hot burned gases will expand and play the role of a piston that greatly accelerates the flame until a transition from deflagration to detonation occurs.

This transition occurs only after the flame has propagated over a minimum distance, known as the "pre-detonation distance", which has the following characteristics:

– it is longer for a less reactive system and for a lower initial flame velocity: too low reactive gases cannot easily detonate, as natural gas (NG);

– it increases with the pipe diameter: for example, it is worth 70 times more than the diameter for propane.

1.2.3.4. *Characteristics of a "rapid deflagration"*

Experience shows that the explosion of an ATEX in the open air can be extremely devastating if the following conditions are met:

– it contains a sufficiently reactive gas other than NG (e.g. hydrogen, or a hydrocarbon such as propane, butane, ethylene, propylene or cyclohexane);

– its volume is very large, at least tens of thousands of m^3;

– it is located in a semi-confined environment, i.e. containing a certain density of structural elements, such as pipes, platforms, tanks, reactors, distillation columns and control rooms (this situation particularly corresponds to chemical or petrochemical plants). The explosion can then result in a phenomenon of "rapid deflagration", the mechanism of which is as follows:

- by propagating, the flame produces very hot burned gases that expand and occupy a much larger volume than the initial ATEX,

- the ATEX fraction that has not yet been reached by the flame undergoes this expansion by flowing into a movement that moves it away from the ignition point and this flow causes the formation of vortices around the surrounding structural elements,

- when the flame reaches these vortices, its surface increases sharply, resulting in its acceleration. Depending on the volume and composition of the ATEX, the flame can accelerate to a speed of 100 m/s or more,

- the explosion produces even more devastating effects than the flame speed is higher.

The flame propagation regime can be qualified as a "rapid deflagration", and the global phenomenon is referred to as VCE (Vapor Cloud Explosion).

1.2.3.5. *Comparison between a detonation and a "rapid deflagration"*

Figure 1.7 shows the comparative profiles of the aerial pressure wave resulting from a detonation and a "rapid deflagration".

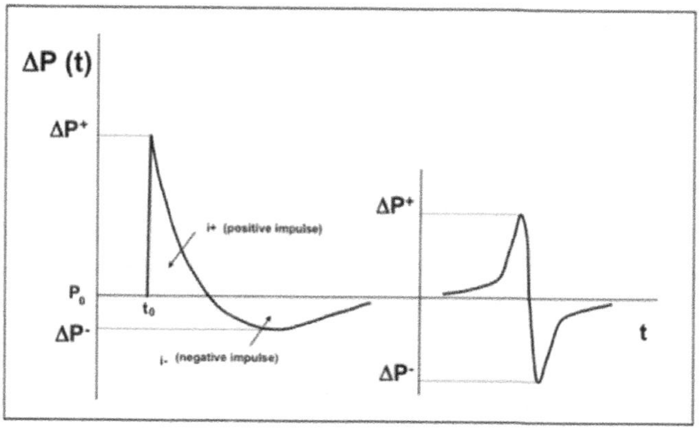

Figure 1.7. *Aerial pressure wave profiles produced by a detonation (left) and a "rapid deflagration" (right)*

There are several differences in these profiles. The characteristics of the wave profile resulting from a detonation are the following:

– at $t = t_0$, the pressure shows a discontinuity which results in a sharp increase of the pressure P(t), from initial atmospheric pressure P_0 ($=P_{atm}$), up to the peak ($P_0 + \Delta P^+$), corresponding to an overpressure of intensity ΔP^+; after this peak, P(t) decreases according to a concave curve, up to a minimum negative value (P_0 - ΔP^-), before increasing again to recover atmospheric pressure;

– the positive phase duration t^+ is significantly shorter than the negative phase t^-;

– the positive impulse i^+ (negative i^-, respectively) is defined as the surface between the P(t) curve and the x-axis for positive P values (negative, respectively).

The profile of the wave from a deflagration is similar to that from a detonation, with the following differences:

– the pressure increases from P_0 to a peak ($P_0 + \Delta P^+$), in a time t^+ much longer than for the detonation;

– the pressure then decreases to the negative value (P_0-ΔP^-) and increases again up to the atmospheric pressure, but in this case, ΔP^- is approximately equal to ΔP^+ (absolute) and the positive phase duration t^+ is comparable to the negative one t^-.

These different profiles explain that the damages caused by the explosion of an ATEX in the "rapid deflagration" regime are not identical to that produced at the same distance and for the same mass of explosive system in the detonation regime. However, it is this rapid explosion mechanism that explains the extent of damages caused by explosions in chemical or petrochemical plants.

1.3. Hollow body rupture (or bursting)

1.3.1. *Definition elements*

When a hollow body (a container) contains a compressed gas, it stores energy in its physical state and, in the event of a rupture (or bursting), this energy is released when the gas expands.

If the container contains a liquefied gas, for example, an LPG such as propane, it adds a phenomenon to that of the expansion of the gaseous phase; it is the phenomenon of BLEVE (Boiling Liquid Expanding Vapor Explosion):

– in the container, the liquid is at a temperature well above its boiling point under atmospheric pressure, i.e. -42°C for propane, for example;

– the difference is even greater if the container is heated (in the case of a container subjected to the flame of an open fire in an LPG storage);

– when the container bursts, the liquid suddenly starts to boil, because the pressure of its vapor is suddenly reduced to atmospheric pressure, and it vaporizes massively;

– as such, the BLEVE phenomenon does not necessarily concern a flammable liquid (water from a water heater may give rise to a BLEVE), but if the container is subjected to the flame of an open fire and if the liquid is a flammable LPG, the vapor released mixes with the air during its dispersion into air and the resulting mixture is ignited by the flame of the fire; the thermal effects are important because due to the radiation of a large thermal flux by the flame. As for the pressure effects, they result from the combustion of the LPG and are then added to the effects of the bursting.

When the liquid is not flammable, the energy released in the explosion is found in effects that are purely mechanical:

– an overpressure wave propagating through air;

– the projection of debris (fragments) from the container.

1.3.2. *Evaluation of E_{lib}*

The energy E_{lib} released in the bursting of a hollow body (in J) can be estimated simply by Brode's relationship [1.34]:

$$E_{lib} = ((P_r - P_{atm}) * V)/(gamma-1) \quad [1.34]$$

where:

– P_r is the pressure of the gas in the hollow body at the rupture time and P_{atm} is the atmospheric pressure (in Pa); strictly speaking, E_{lib} is proportional to $(P_r - P_{atm})$, but it can generally be considered that $P_r >> P_{atm}$, so that E_{lib} is proportional to P_r;

– V is the volume of the hollow body (in m^3);

– γ (gamma) is the ratio of the thermal mass capacities of the gas contained in the hollow body ($\gamma = c_p/c_v$).

To estimate E_{lib}, it is necessary to be able to estimate P_r and, in the case of simple geometric forms such as a cylinder, this estimation is relatively easy, with Lamé relationship [1.35]:

$$P_r = 2\ e\ \sigma_{tang}/D = 2\ e\ \sigma_r/D = \sigma_r\ e/R \quad [1.35]$$

where:

– e is the cylinder thickness (in mm);

– D is the diameter of the cylinder (in mm);

– P_r is the cylinder rupture pressure (in Pa);

– σ_{tang} and σ_r are the tangent and radial rupture stresses of the hollow body material (MPa), respectively.

During the investigation of the accidental bursting of a 40 L steel bottle (see case study 25), the energy released in the bursting, estimated from Lamé and Brode relationships (the reader is referred to this case study for the establishment of Lamé's relationship), has to be compared with the energy necessary for the production of the observed damage.

1.4. Superheated liquid vaporization

1.4.1. *Definition elements*

An explosive phenomenon occurs when a liquid heated to a temperature T_L is put in contact with a hot body, heated to a temperature T_B such as $T_B >> T_L$. If T_B is much higher to the boiling temperature T_{eb} of the liquid, this contact causes the sudden vaporization of the liquid.

The explosive system consists of the combination of the liquid and the hot body and stores energy in the physical state.

Such a system can meet:

– in metallurgical industries such as foundries (steel or other metals), where water is a fluid, frequently used to cool certain elements of the facility, which would be degraded by too high of a temperature;

– in boilers (rupture of a water tube near the combustion chamber).

The energy released in the explosion is found in the effects that are purely mechanical:

– an aerial overpressure wave propagating through the air;

– the potential projection of debris.

1.4.2. *Evaluation of E_{lib}*

An explosion resulting from the sudden vaporization of a liquid in contact with a hot body implies a change of state of the liquid and involves a maximum energy E_{lib} equal to (relationship [1.36]):

$$E_{lib} = m\,(c_p\,\Delta T + \Delta H_v) \qquad [1.36]$$

where:

– m is the mass of the vaporized liquid (in kg);

– c_p is the thermal mass capacity of the liquid (in J/kg.K);

– ΔT is the temperature difference between the cold liquid and the warm body (in K);

– ΔH_v is the enthalpy of liquid vaporization (in J/kg).

In reality, the effects of the explosion result from the release of a much lower energy than E_{lib}, because the heat transfer that occurs between the hot body and the vaporizing liquid involves their contact surface to be affected rather than their volume. Although the efficiency of this transfer is higher if the hot body is itself in the state of molten liquid, it is still well below 1.

1.5. Comparison of E_{lib} with the energy E_{eff} required to produce the explosion effects

E_{eff} is the energy needed to produce the explosion effects.

For each of the four types of explosion, it is expected that the E_{lib} value calculated using equations [1.16] [1.33] [1.34] and [1.36] will be greater than the E_{eff} value needed to produce the explosion effects.

The causes of this difference include:

– for explosive systems 1 (ATEX) and 2 (exothermic chemical reaction), which combine several components, the composition of the system is never completely homogeneous and never corresponds to the stoichiometric composition;

– for system 4, the heat transfer involves the contact surface between the hot body and the vaporizing liquid and is more dependent on the value of this surface than on the volume of the hot body;

– moreover, unlike these systems, system 3 (hollow body bursting) involves a phenomenon for which the concept of composition is absent;

– a fraction of the energy released E_{lib} is found both in the form of working friction forces and pressure losses, and also in thermal losses, especially by the radiation effect of the flame.

This means that any explosion is characterized by the existence of a yield ρ, which is always less than 1 (relationship [1.37]):

$$\rho = E_{eff}/E_{lib} < 1 \qquad [1.37]$$

For systems (1), (2) and (4), the value of ρ may be much less than 1. On the contrary, the value of ρ for system 3 is very close to 1.

2

Method of Investigating an Explosion

2.1. Introduction

As indicated in the introduction, the investigation of an explosion aims to establish its mechanism as fully as possible, in order to determine all its elementary causes.

This initial work is necessary for subsequent important phases, which consist of:

– identifying appropriate measures to prevent the occurrence of a new disaster, either in the damaged facility or in similar facilities, and also to protect against the effects of those that may occur, despite the measures taken;

– using the feedback of the explosion and generalizing these measures, translating them into best practices and possibly regulatory constraints.

2.2. Establishment of the explosion mechanism

To establish the mechanism of an explosion consists of answering the following three questions:

– WHAT? What is the nature of the explosive system that exploded?

– HOW MUCH? How much explosive system was involved in the explosion?

– HOW? How did the explosive system appear and how has its explosion been triggered?

This chapter presents a method that allows access to the mechanism of explosions, successively answering the following questions:

1) first, answer the HOW MUCH question, considering the effects (damage) that the explosion produced and determining the energy released;

2) the crossing of the amount of energy released with the explosion circumstances enables us to identify the nature of the explosive system and thus to answer the question WHAT?;

3) finally, the knowledge of the nature of the explosive system, as well as the circumstances of the explosion, generally provide an answer to the question HOW?

2.3. Search for answers to the questions of HOW MUCH and WHAT

The search for answers to the questions HOW MUCH? and WHAT? goes through the following steps:

– a list of the different types of damage caused by the explosion;

– an estimate of the energy E_{eff} required for their production;

– a hypothesis on the type of explosive system involved, among the four presented in Chapter 1 (see sections 1.1, 1.2, 1.3 and 1.4), taking into account the circumstances of the explosion (products used, processes operated, etc.) and the value of E_{eff}:

1) the explosive system is an EXplosive ATmosphere (ATEX),

2) the explosive system is a mixture of substances which may give rise to a highly exothermic chemical reaction and may be explosive under certain conditions (it may also be a single substance which is capable of an explosive decomposition),

3) the explosive system is a hollow body which bursts under the pressure of its contents, being in the gaseous state and possibly also in the liquid state,

4) the explosive system consists of a hot body that comes into contact with a cold liquid and causes it to vaporize abruptly;

– an E_{lib} estimate taking into account E_{eff} and the yield associated with the type of explosion envisaged;

– an estimate of the quantity of the explosive system involved, according to the E_{lib} value corresponding to the type of explosion envisaged:

- type 1: mass of the flammable product involved in the ATEX explosion,

- type 2: mass of the chemical system other than an ATEX involving an exothermic reaction between two or more reagents or an exothermic reaction of a single reagent,

- type 3: characteristics of the volume and rupture pressure of the hollow body involved,

- type 4: mass of the cold liquid brought into contact with the hot body and difference in temperature between the cold liquid and the hot body;

– a comparison of the quantity of the explosive system deduced from the E_{lib} value with that compatible with the circumstances of the explosion.

The compatibility of these two quantities constitutes a validation element of the hypothesis made in step 3; on the contrary, the hypothesis is not verified if the quantity of the explosive system deduced from the E_{lib} value is significantly greater than that corresponding to the circumstances of the explosion.

2.4. Identification of the different types of damage produced by an explosion

When we get to the site where an explosion occurred, what we find is the damage it caused.

An explosion is characterized by the production of more or less significant damage, both to property and to people. The term “damage” may be considered to be synonymous with the term “effects” or the term “adverse effects”, although the use of the term “effects” gives this term an attenuated meaning in relation to the term “damage”.

We will henceforth use either the word “damage” or “effect” to describe the damage caused by an explosion, without the term “effect” referring to damage of less intensity than the term “damage”.

The establishment of the mechanism of an explosion begins with the identification of the effects, both in nature and intensity, that the explosion has produced, knowing that:

– the damage is mechanical and possibly thermal;

– it may have been produced:

- on structures (and the environment in the broad sense: soil or plants, for example) which suffer mechanical and possibly thermal effects,

- on persons who suffer from physiological effects related to exposure to overpressure and/or heat flux or elevated temperature for a period of time.

2.4.1. *Effects on structures*

2.4.1.1. *Mechanical effects*

The characterization of the mechanical effects of an explosion requires us to define the concept of the "center of the explosion".

An explosion releases its energy E_{lib} in a more or less confined area of space:

– a confined space is, for example, inside a process capacity or equipment, or inside the room of a housing place;

– a semi-confined space is, for example, the immediate environment of a refinery or chemical plant, possibly congested by pipe racks, industrial equipment or buildings;

– a space not covered is, for example, the bare ground above an underground pipeline or the distant environment of an offshore platform.

The center of the explosion can be defined as the location where the explosive system was concentrated prior to the initiation of the explosion.

From the point of view of the damage produced, it turns out that:

– the center of the explosion is where they are most important;

– the intensity of the damage decreases as the distance R to the center of the explosion increases;

– the aerial wave produced propagates through the air from this center, according to a radial symmetry a priori, except for special reasons taking into account the presence of obstacles on which the wave reflects: this reflection affects both the intensity and the direction of propagation of the reflected wave and thus also the intensity of the damage it produces;

– debris or fragments are projected from this center, also in a radial symmetry, except for special reasons that take into account the presence of obstacles on which they may ricochet.

The mechanical effects result from the exercise of pressure forces that develop:

– during the expansion of the combustion fumes produced in the flame (Explosive Systems 1 and 2);

– when the gas initially contained in the hollow body is released (Explosive System 3);

– during the expansion of the vapor emitted by the boiling liquid in contact with the hot body (Explosive System 4);

– in some cases, these different effects can be combined;

– they take different forms depending on the type of explosion and the degree of containment of the space in which it occurs:

- If the explosion of an ATEX (explosive system 1) occurs in the confined space of a container, the pressure developed is exerted on its walls which can either withstand the maximum overpressure reached, either with or without permanent deformation, or rupture locally, or totally. The example of buildings is dealt with in detail in Part 2, in cases of domestic explosions.

- In the case of an explosion occurring in open air or in a confined space, the pressure forces result from the action of an aerial overpressure wave which gradually attenuates until it becomes a sound wave and which is characterized by a more or less rapid rise front depending on the type of explosion considered and on the mass of the explosive system involved. This situation is dealt with in Part 2, in the case of accidental explosions in industrial petrochemical and chemical installations.

- Finally, when the explosion causes the rupture of structural elements, it generally also causes the projection of debris or fragments from these elements: this is particularly true in the case of the bursting of a hollow body (Explosive System 3), which will be the subject of the case 25, dealt with in Part 2.

For the effects of overpressure on structures, the legislation generally retains the threshold values given in Table 2.1.

Type of damage	**Overpressure threshold (mbar)**
Significant glass breakage (more than 10%)	20
Minor damage	50
Severe damage	140
Domino effect	200
Very severe damage	300

Table 2.1. *Thresholds for overpressure effects on structures*

2.4.1.2. *Thermal effects*

The thermal effects of an explosion result from the production of a flame and therefore they only appear in type 1 and type 2 explosions.

2.4.1.2.1. Convection effects

Some combustible materials may ignite or at least be marked by the flame when they are in direct contact with it (usually low-density and low-thermal inertia, such as paper, wood or plastics in small thicknesses). A flammable liquid having a low flash point ignites on contact with the flame of an explosion.

Strictly speaking, the intensity of the thermal effects depends on the duration of the contact between the combustible body and the flame: therefore, in the case where it propagates in the detonation regime, it is far too brief to produce the least thermal effect.

This difference can be used to establish that the explosion of an ATEX occurred in a deflagration regime rather than a detonation one.

2.4.1.2.2. Radiation effects

The intensity of the effects produced by the flame's radiation depends on its surface and its distance from the bodies subjected to its radiation. But it depends above all on its duration and, in general, its brevity is decisive and its effects are negligible.

In the case of the BLEVE of a propane tank caught in a fire, which can be considered as an Explosive System 3 to account for the initial burst of the tank, the resulting combustion of propane is an epiphenomenon with a flame that is both durable and very large, its radiation producing intense thermal effects, very devastating to facilities and deadly to people.

2.4.2. *Effects on the human body*

2.4.2.1. *Mechanical effects*

An explosion produces different types of physiological effects on the human body, the nature and severity of which depend on the level of overpressure reached.

For overpressure effects, the legislation generally retains the threshold values given in Table 2.2.

Types of effects produced (nature and intensity)	Overpressure threshold (mbar)
Injuries by broken glass (indirect effects)	20
Irreversible injuries (in particular, the possibility for a person in a standing position to be knocked over and thrown violently onto the ground or against a rigid structural element, which exposes them to the risk of a potentially fatal skull fracture)	50
First lethal effects (area of serious danger to human life)	140
Significant lethal effects (area of very serious danger to human life)	200
Damage to eardrums	300
Potentially fatal lung injury (hemorrhage)	500

Table 2.2. *Threshold for the effects of overpressure on the human body*

2.4.2.2. *Thermal effects*

The flame of an ATEX explosion is hot enough to seriously burn any skin surface in direct contact with it and a person who has been surrounded by the flame may be fatally burned.

In the case of the BLEVE mentioned above, the flame radiation causes fatal burns up to high distances (see section 2.4.1.2.2).

2.5. Estimation of the energy required to produce the mechanical effects

It is through an analysis of the nature and intensity of the damage that the establishment of the explosion mechanism begins:

– the nature of the damage is related to the mechanism of the explosion and therefore to the nature of the explosive system involved;

– the more energy this explosive system releases, the more intense the damage.

In the following, we call E_{eff} the estimated value of the energy that was needed to produce the observed damage.

2.5.1. *E_{eff} estimation tools*

Several tools are available to estimate E_{eff}

2.5.1.1. *E_{eff} estimate using TNT equivalent*

To analyze the damage caused by an explosion, the concept of TNT equivalent is commonly used.

TNT or trinitrotoluene is a chemical compound with the formula $C_7H_5N_3O_6$.

It is likely to give rise to explosive decomposition (Explosive System 2) with the following characteristics:

– the reaction is highly exothermic (ΔH = - 4.69 MJ/kg);

– at flame temperature, decomposition products are all gases: CO_2, CO, H_2O, N_2, NO_2;

– under certain conditions, its decomposition occurs in detonation regime.

TNT has long been used as an explosive substance, for civilian or military purposes and in particular as a component of explosive mixtures. It is in this use that it has been acquired a very good experience of the nature and intensity of damage that the explosion of a mass of TNT produces on a target, depending on the distance between that mass and the target and whether the charge is placed on the ground or at a distance from the ground.

2.5.1.1.1. Definition of Eq_{TNT} equivalent of an explosion

The TNT equivalent of an explosion, Eq_{TNT}, is defined as the mass m of TNT whose explosion would have produced the same damage, at the same distance from the center of the explosion.

2.5.1.1.2. Eq_{TNT} as energy equivalent

EqTNT can be considered as a simple energy equivalent and it is, for example, possible to determine the TNT equivalent of a 45 g chocolate bar (Figure 2.1):

– on the packaging, it is indicated that it has an energy equivalent of 982 kJ;

– this means that this chocolate bar has a TNT equivalent of about 200 g.

Figure 2.1. *View of a 45 g chocolate bar. For a color version of this figure, see www.iste.co.uk/chaineaux/mechanisms.zip*

However, the reactions of chocolate digestion and TNT decomposition are very different:

– the durations of each phenomenon are very different:

- the detonation of 1 kg TNT lasts less than 0.0001 second;

- the concept of AJR (Recommended Daily Intakes) enables us to estimate at 24 h the time of assimilation of a food energy, i.e. 8,360 kJ;

– the powers released by each phenomenon are therefore also very different:

- a power of 1 W being equal to an energy of 1 J released in 1 s, the power released during the explosion of 1 kg of TNT is therefore worth 46.9 TW;

- whereas the assimilation of AJR corresponds to a power of only 100 W (this figure corresponds to the order of magnitude of the power radiated by the human body).

The comparison of the chocolate bar and the TNT stops here.

On the contrary, different types of accidental explosions are comparable to the TNT explosion, by the nature and intensity of the damage they produce and it is to make this comparison that it is interesting to determine their TNT equivalent.

2.5.1.1.3. Determining the TNT equivalent of an explosion

An explosion can only be characterized by its TNT equivalent if it has produced an air overpressure wave: indeed, regardless of the fact that it could be accompanied by the projection of debris or fragments, it is this air wave that, by propagating into space from the center of the explosion, produces damage whose intensity decreases as the distance R to the center of the explosion increases.

The determination of the TNT equivalent of an explosion is based on the use of the abacus TM5-1300 (US Department of Navy 1990) of Figure 2.2, which links the peak overpressure value of the aerial wave (or maximum overpressure) ΔP, produced at the distance R from the center of the explosion, by the detonation of a mass m of TNT, laid on the ground, at the size λ, called "reduced distance", which is defined by the relation $\lambda = R/(m)^{1/3}$.

The TNT equivalent determination process consists of the following five steps:

1) draw a damage map, to identify each type of damage produced, and to measure the distance R between the location where it was produced and the center of the explosion;

2) allocate, for each type of damage, the peak overpressure ΔP of the wave to the location of the damage;

3) transfer the value of ΔP to the abacus of Figure 2.2;

4) note the corresponding value of λ;

5) calculate the mass m of TNT that would have produced this type of damage at distance R, from the relationship [2.1]:

$$m = (R/\lambda)^3 \quad [2.1]$$

For example, the threshold for breaking glass, about 20 mbar, corresponds to $\lambda = 70$ m/(kg)$^{1/3}$: an explosion having a TNT equivalent equal to10 kg generates an aerial overpressure wave capable of breaking glass at the distance R of about 150 m.

2.5.1.1.4. E_{eff} determination by using TNT equivalent

To determine the energy E_{eff} (in MJ) that produced the observed damage, simply multiply the value of Eq_{TNT} by the decomposition energy of the TNT (relationship [2.2]):

$$E_{eff} = 4.69 \times Eq_{TNT} \text{ (in MJ)} \quad [2.2]$$

Since it is possible to estimate the value of E_{eff}, the determination of the TNT equivalent of an explosion is a necessary first step in establishing its mechanism.

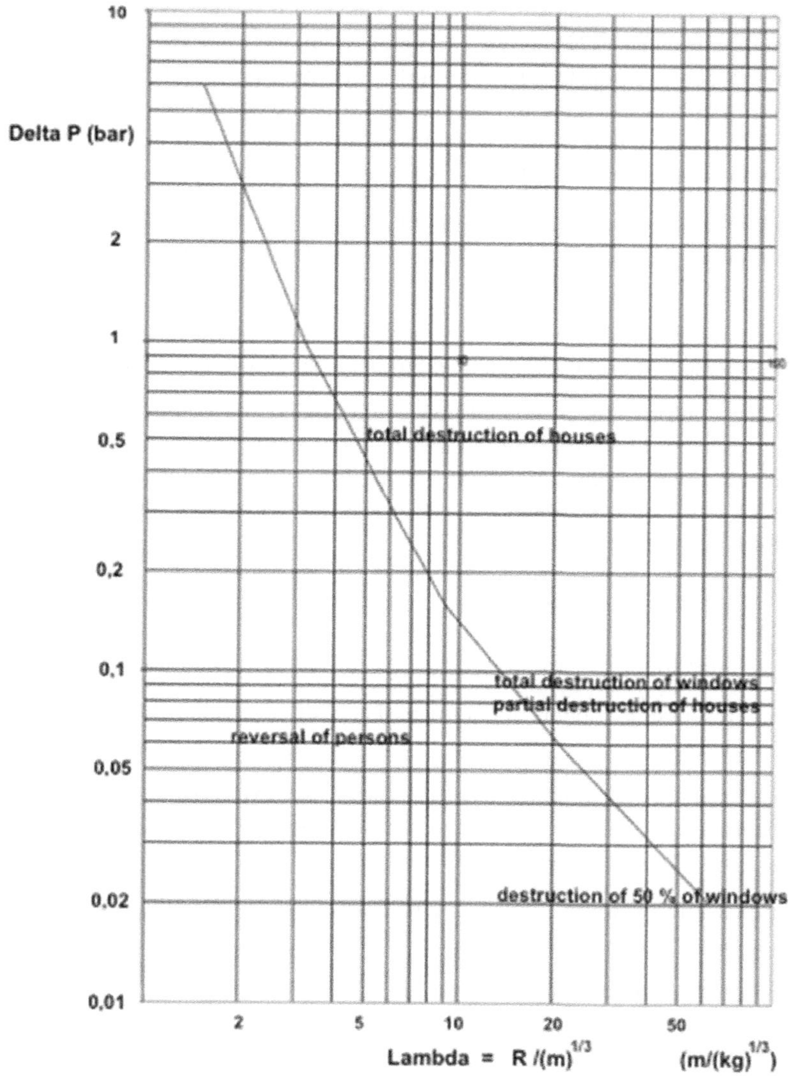

Figure 2.2. *Extrapolated TM5-1300 abacus, giving the overpressures ΔP caused by the detonation of a mass of TNT laid on the ground*

2.5.1.1.5. TNT equivalent model limitations

The use of the TNT equivalent model is suitable for any explosion which results in the emission of a shock wave:

– this is the case of Explosive System 2 which corresponds to the decomposition of a single substance (TNT is an example) or to the very exothermic chemical reaction between two substances; the solid ammonium nitrate (NH_4NO_3) is another example of a substance likely to detonate and it is the detonation of a few hundred tons of this product that is at the origin of the explosion at AZF's facilities in Toulouse on September 21, 2001, or in Beirut in August 2020;

– another example is Explosive System 3, where the rupture of a hollow body is accompanied by the sudden release of the gas it contained, producing a shock wave very similar to that of a detonation;

– Explosive System 4, where the vaporization of a liquid in contact with a hot body takes place in a brutal enough manner that a shock wave is emitted;

– on the contrary, in an unconfined environment, the explosion of an ATEX (Explosive System 1) does not necessarily give rise to the emission of a shock wave, because the flame does not propagate in the detonation regime but in the deflagration one, which is quite different. A priori, the model of the TNT equivalent is therefore not suitable in this case.

2.5.1.2. *E_{eff} estimation using other models*

2.5.1.2.1. Introduction to models other than TNT equivalent

Before discussing models other than the TNT equivalent, the following are presented successively:

– the differences between the two possible ATEX explosion regimes, detonation and deflagration, in terms of their mechanism and also the characteristics of the aerial overpressure wave they produce;

– the possible evolution, under certain conditions, of a deflagration towards a detonation or towards a "rapid" deflagration;

– characteristics of a rapid deflagration.

2.5.1.2.2. Presentation of the multi-energy method

To overcome the imprecision of the Eq_{TNT} model when it is used to estimate the energy released in a rapid deflagration, various tools have been developed, such as the multi-energy method that takes into account the level of congestion of the ATEX where the explosion propagated and thus also the speed of the flame.

This method, developed by TNO (TNO Multi Energy Method 1996), makes it possible to estimate the peak overpressure ΔP of the aerial wave that would be produced at the distance R by the explosion of an ATEX of stoichiometric composition, containing a mass m of a given fuel.

2.5.1.2.3. Application of the multi-energy method

The use of the abacus in Figure 2.3 allows a graphical application of the method.

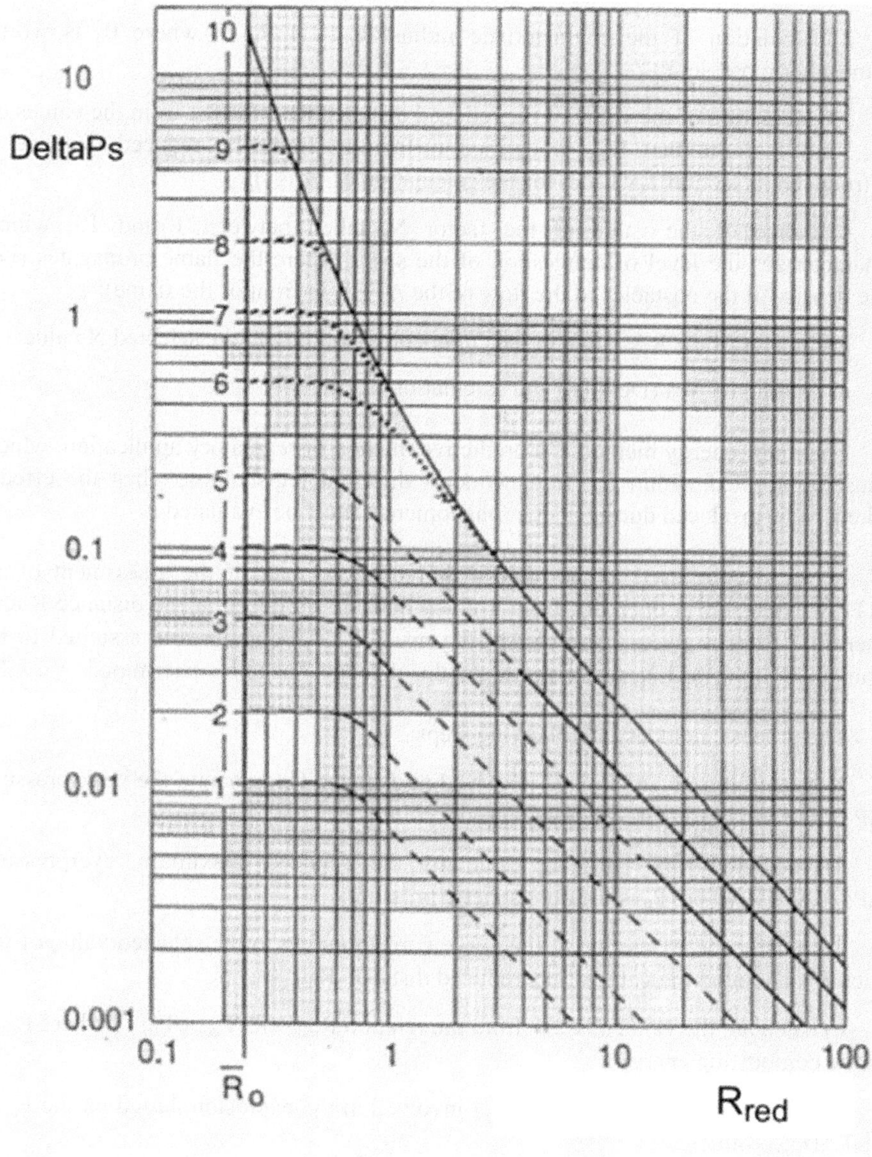

Figure 2.3. *Abacus used for a graphical application of the multi-energy method*

The process includes the following steps:

1) calculation of the combustion energy E_{comb} that would be released in the explosion, from m and the fuel combustion enthalpy ΔH_{comb};

2) calculation of the characteristic radius $R_c = (E/P_0)^{1/3}$, where P_0 is worth atmospheric pressure;

3) calculation of the value of the reduced distance $R_{red} = R/R_c$, from the values of Rc, and R (note that R_{red} is defined differently from the reduced distance λ introduced in section 2.5.1.1.3, for the calculation of Eq_{TNT});

4) choice of the value of the factor N, taken between 1 and 10, which characterizes the level of congestion of the space where the flame propagates (i.e. the density of the obstacles to the flow of the ATEX in front of the flame);

5) transfer of the R_{red} value to the curve corresponding to the selected N value;

6) reading of ΔPs (DeltaPs) and calculation of ΔP.

The multi-energy method is a predictive method of regulatory application, which must be applied within the framework of different legislations, when the effects likely to be produced during a VCE phenomenon must be evaluated.

It can also be used in the opposite direction: as part of the assessment of an ATEX explosion in open air, from a given damage produced at the distance R and therefore from a given value of ΔP, the mass m of a combustible assumed to be contained in the ATEX, which produced this damage, has to be determined.

The process includes the following steps:

1) based on a given damage observed at distance R, associate the peak pressure ΔP of the aerial pressure wave that produced this damage;

2) calculate from ΔP the value of the reduced maximum overpressure $\Delta Ps = \Delta P/P_0$, where P_0 is the atmospheric pressure;

3) transfer the ΔPs value to the curve corresponding to the selected value of the factor N and read the value of the reduced distance R_{red};

4) calculate the value of E_{comb} from the relation $R_{red} = R/(E_{comb}/P_0)^{1/3}$, where E_{comb} is the combustion energy;

5) calculate the mass m of the fuel involved in the explosion, based on the E_{comb} and ΔH_{comb} values.

Example:

– for a damage of glass breakage assumed to correspond to $\Delta P = 2{,}000$ Pa, the calculation gives $\Delta Ps = \Delta P/P_0 = 2{,}000/100{,}000 = 0{,}02$;

– the transfer of this ΔPs value to the curve corresponding to N = 5 (a value commonly used for medium congestion) leads to the value of the reduced distance $R_{red} = 7$;

– from this value of R_{red} and R = 150 m, the calculation leads to the value of E_{comb}, which is approximately 1,000 MJ;

– based on the E_{comb} value and assuming that the gas is propane, whose combustion heat ΔH_{comb} is 50 MJ/kg, the calculation leads to the value m = 20 kg.

2.5.1.3. *Other methods*

There are other methods for estimating the peak overpressure ΔP of the aerial wave that would be produced at the distance R by the explosion of a stoichiometric composition ATEX, containing a mass m of a given fuel.

One of them is the BAKER method (Baker 1979) which is similar to the multi-energy method, in that it takes into account the degree of congestion of the space where the flame of the explosion propagates.

2.5.1.4. *Choice of the best fit model*

In order to evaluate the quantity of an explosive system of type 1, 2, 3 or 4 involved in the explosion, based on the damage it has produced, it is necessary to choose the most relevant model.

For each explosive system, Table 2.3 gives the main characteristics of the phenomena at the origin of the explosion and the models to be used.

The table shows that:

– the TNT equivalent model is the only one applicable when estimating the quantity of a type 2, 3 or 4 explosive system from the damage caused by the aerial pressure wave resulting from its explosion;

– in the case of a VCE, the multi-energy method gives a more reliable result than the TNT equivalent model and must be used;

– in the case of a type 1 explosive system where the ATEX explosion occurs in open air, modeling is not necessary if the ATEX has a limited volume and/or its composition is close to the explosion limit: the flame speed will remain low, as well as the intensity of the aerial pressure wave and the damage produced will be very low.

Type of explosive system	Main characteristics of the phenomenon involved and the explosive system	Flame propagation regime	Model to be used
1	Unconfined ATEX of limited volume	Deflagration regime, without emission of an intense air overpressure wave	No model required No model required
	VCE (High Volume ATEX)	Rapid deflagration regime with intense aerial overpressure wave emission	Multi-energy method
2	Initial system of small volume, solid or liquid	Detonation or deflagration of the system, with emission of an aerial overpressure wave	TNT equivalent
3	Burst of a hollow body	Neither combustion nor flame Emission of an aerial overpressure wave	TNT equivalent
4	Sudden vaporization of a liquid	Neither combustion nor flame Emission of an aerial overpressure wave	TNT equivalent

Table 2.3. *Identification of the model to be used for each explosive system defined by the characteristics of the phenomenon involved*

2.5.2. Comparison between E_{eff} and E_{lib}

After estimating E_{eff}, it is necessary to estimate the value of E_{lib}, i.e. the energy that was released during the explosion. It would be logical to have (relationship [2.3]):

$$E_{lib} = E_{eff} \quad [2.3]$$

In reality, the amount of energy E_{lib} that was released by the explosion is necessarily greater than the amount of energy E_{eff} that was used to produce the damage, according to the following inequality [2.4]:

$$E_{eff} < E_{lib} \quad [2.4]$$

The reason for this inequality is a consequence of the two fundamental principles that govern physics:

– a first fundamental principle is the principle of energy conservation: in all energy transfers, the energy quantity is conserved;

– but a second fundamental principle that immediately follows from the first and corrects it is the principle of the degradation of energy: in reality, energy is not entirely conserved and it degrades during the transfer, mainly due to heat losses in thermal processes and friction in mechanical processes.

E_{lib} being higher than E_{eff}, the notion of the yield ρ of an explosion must be introduced, which corresponds to the following definition (relationship [2.5]):

$$E_{eff}/E_{lib} = \rho < 1 \qquad [2.5]$$

2.5.3. *Order of magnitude of the yield ρ for each type of explosive system*

The values of ρ depend on each type of explosive system:

– in the case of ATEX explosions in unconfined or semi-confined environments, the yield ρ is significantly less than 1:

- about 30 accidental VCE-type hydrocarbon explosions (i.e. involving rapid deflagration), where the mass of hydrocarbon discharged into air was known, were analyzed and it was found that the average value of ρ was only a few percent and never exceeded 10% (Lannoy 1984),

- this low value is mainly due to the fact that only a small proportion of the hydrocarbon discharged is involved in the formation of ATEX;

– for a reaction between two products, ρ is clearly less than 1 since, in the conditions of an accident, their proportions are not stoichiometric and the mixture is not homogeneous (it is a little higher for the decomposition of a single product since the notion of homogeneity of the mixture does not exist);

– in contrast, for the bursting of a hollow body, the yield is very close to 1 (relationship [2.6]):

$$E_{lib} \approx E_{eff} \qquad [2.6]$$

– finally, for explosions resulting from the vaporization of a liquid in contact with a hot body, ρ is clearly less than 1, because the transfer of energy takes place at the contact surface (however, the efficiency is greater if the hot body is also liquid).

2.6. Hypothesis on the type of explosion involved

We have just seen that the type of explosive system involved has an influence on:

– E_{lib} (cf equations [1.16], [1.32], [1.35] and [1.36]);

– the yield ρ of the explosion that links the E_{lib} and E_{eff} magnitudes by the following inequality [2.7]:

$$E_{lib} > E_{eff}/\rho, \text{ with } \rho < 1 \quad [2.7]$$

At this stage of the investigation, it is therefore necessary to make a hypothesis about the type of explosive system involved in the explosion.

It is both the analysis of the circumstances and the effects of the explosion that make it possible to formulate this hypothesis:

– the nature of the processes used;

– the flammability or chemical reactivity of the products used, either alone or in a mixture;

– information from feedback and known accidents involving similar processes or products.

2.7. Estimation of the quantity of the explosive system involved

2.7.1. *General case*

Depending on the hypothesis on the type of explosion involved and on the corresponding value of ρ, it is possible to estimate E_{lib} from the values of E_{eff} and thus also the quantity of the explosive system, namely:

– type 1: mass of the fuel involved in the ATEX explosion;

– type 2: mass of the chemical system other than an ATEX, having involved an exothermic reaction of decomposition of a single reagent, or an exothermic reaction between two or more reagents;

– type 3: characteristics of the volume and pressure of the hollow body involved;

– type 4: mass of the cold liquid in contact with the hot body.

2.7.2. *Specific case of an ATEX explosion occurring in a confined environment*

The case of an explosion of an ATEX (Explosive System 1) occurring in a confined environment, i.e. inside a "container" (this word must be taken in the broad sense because it may be a building, or even a living space) is particular for the following reasons:

– the analysis of the damage intensity does not usually enable us to determine accurately the mass of the explosive system involved;

– but, in this case, it appears that knowledge of this characteristic is not really necessary to determine the explosion mechanism.

The reason for this is double:

– the mechanical strength of the ATEX container is generally very low, so that a small mass of the combustible product is sufficient to explain the damage that the container has suffered and that can go as far as its destruction (this mass can therefore be contained in an ATEX which occupies only a small fraction of its volume);

– the determination of the mechanism of the explosion involves the identification of the explosive system. However, it is common for the Explosive System 1 to impose itself in view of the circumstances of the explosion (possibility of a leak of a flammable gas implemented in a building and absence of any other explosive system, etc.). In this case, the determination of the mechanism of explosion is more concerned with the search for the conditions of formation and ignition of ATEX (origin of the gas leak, duration of formation of ATEX, nature and location of the ignition source, etc.).

2.8. Evaluation of the hypothesis on the type of explosion involved

2.8.1. *Compatibility of the hypothesis with the circumstances of the explosion*

To evaluate the hypothesis on the type of explosion involved, it is necessary to compare the quantity of the explosive system deduced from the E_{lib} value with that which is compatible with the circumstances of the explosion:

– the compatibility of these two quantities constitutes a validation element of the hypothesis made in step 3;

– on the contrary, if the quantity of the explosive system deduced from the E_{lib} value is much higher than that corresponding to the circumstances of the explosion, the hypothesis is not verified and another hypothesis must be envisaged;

– it is possible that this other hypothesis should lead to the possibility of another explosive system, which is also compatible with the circumstances of the explosion.

2.8.2. *Cases of explosions which may a priori involve different systems*

2.8.2.1. *Case of a boiler explosion*

When an explosion occurs in a water-tube boiler facility, supplied by a flammable gas or liquid, three mechanisms can be envisaged a priori:

– first mechanism: a sudden vaporization of water discharged into the combustion chamber, in case of the puncture of a water tube;

– second mechanism: explosion of an ATEX formed in the combustion chamber, following a deregulation of the composition of the mixture supplying the burner;

– third mechanism: explosion corresponding to Explosive System 3 and involving a tank under air pressure, associated with the boiler.

It happened that, in various accidental explosions, either one of these three mechanisms was definitely involved.

A last mechanism can be envisaged a priori: it would be a consequence of the former insofar as it assumes that, when water is brought to a very high temperature in contact with a very hot body, it can undergo thermolysis, i.e. a thermal decomposition into hydrogen and oxygen, which could then be followed by the formation and explosion of a hydrogen–oxygen ATEX. In fact, for reasons related to both thermochemistry and chemical kinetics, such a mechanism has no physical reality, under the conditions of the introduction of water into a combustion chamber, for the following reasons:

– water thermolysis is a highly endothermic phenomenon (the energy required for it to occur is the opposite of the hydrogen combustion energy in oxygen);

– the kinetics of the reaction are significant only at very high temperatures and are greatly slowed by cooling of the medium.

Finally, the possible mechanism of an explosion involving a gas boiler is either:

– a physical explosion by sudden vaporization of water (Explosive System 4);

– a physical explosion by bursting a hollow body under air pressure up to a few dozen bars (Explosive System 3);

– the explosion of a confined ATEX, formed in the combustion chamber (Explosive System 1).

In order to definitively validate the hypothesis relative to the nature of the explosive system, it must be verified, on the one hand, that the nature of the damage is compatible with the retained scenario and, on the other hand, that the amount of system required to produce the observed damage was actually available under the conditions of the explosion.

In Part 2, case 27 is relative to an explosion in a boiler.

2.8.2.2. *Other examples*

Part 2 is devoted to the presentation of 27 case studies of accidental explosions. For some of them, several of the four explosive systems could have been involved a priori. For the following example cases, the system actually involved and the explosion mechanism have been identified by applying the method described in Chapter 2:

– Explosive System 1: case study 7 (explosion in a pulp paper tank);

– Explosive System 2: case study 20 (explosion in an aluminum foundry);

– Explosive System 3: case study 25 (cylinder burst);

– Explosive System 4: case study 27 (explosion in a household waste incinerator).

2.8.2.3. *Comparison of characteristics of significant explosions*

For five explosions that occurred in France in recent decades, some of which have been particularly devastating and which can be considered representative of each of the Explosive Systems 1 to 4, their main characteristics are given in Table 2.4:

– the nature of the explosive system;

– the quantity of the explosive system (its mass m for Systems 1, 2 and 4, or the burst pressure P_r and the volume V of a hollow body for System 3);

– the EqTNT;

– the explosion yield ρ.

Four of these five explosions are some of the cases dealt with in Part 2, which were examined.

By way of comparison, the last line of the table contains information from the literature on the atomic explosion of the bomb that was dropped on Hiroshima on 6 August, 1945.

Reference of explosion	Nature of explosive system and phenomenon	Mass m of explosive system or bursting pressure P_r and volume V of hollow body	TNT Equivalent EqTNT (kg)	Explosion yield ρ
VCE at refinery – La Mède (France)	1 Rapid deflagration of an ATEX air – butane	5 t (ATEX air-butane)	5,000	0.1
Explosion in an aluminum foundry	2 Reaction (liquid Al + H_2O)	30 to 90 kg of liquid aluminum	200 to 500	≈ 0.2
Explosion at the AZF plant (Toulouse)	2 Detonation of NH_4NO_3 (ammonium nitrate)	200 to 400 t of a heap of ammonium nitrate	20,000 to 40,000	0.1
Explosion in a filling cylinder center	3 Burst of a 40 L cylinder containing chlorine	P_r = 500 bar V = 0.04 m^3	1	≈ 1
Explosion in a steel foundry	4 Sudden vaporization of water in contact with molten slag	Vaporization of water in contact with about 10 kg of liquid slag at about 1,000°C, contained in a solid slag mass	1	<< 1
Hiroshima bomb		60 kg uranium	15,000,000	0.01

Table 2.4. *Characteristics of remarkable accidental explosions corresponding to one of the four types of explosive system*

2.9. Search for answers to the question of HOW?

After identifying the nature of the explosive system involved (question of WHAT?) and assessing the quantity involved (question of HOW MUCH?), the last question to be answered is HOW?

To answer this question, all available information must be gathered on the circumstances in which the explosive system appeared and the explosion was triggered.

In the case of an industrial facility, this information will of course have to be collected from the operator. Specifically, they concern:

– the operating conditions of the process involved (values of the operating parameters measured/recorded at the time of explosion, as well as the nominal values of those parameters);

– the particular circumstances prevailing at the time of the explosion (e.g. technical and/or human failures);

– the testimonies gathered on how the explosion was perceived, while taking into account the fact that human testimonies are subjective and therefore not certainly reliable;

– exploitation of the feedback from similar explosions in terms of effects, circumstances or mechanism.

2.10. Representation of the mechanism of explosion by tree of events

The event tree method is a frequently used accident analysis method to identify and understand its causes:

– it consists of building an organization chart that lists all the basic events that were necessary for the occurrence of the explosion;

– each basic event is inserted into the organization chart, being at the same time:

- the consequence of one or more other events,

- the cause of one or more other events.

This method is similar to the tree of causes method, developed by the INRS (Institut National de Recherche et de Sécurité 2019). This is a structured and rigorous method for determining the accidental mechanism on the basis of facts, and is based on a participatory approach in working groups, with the objective of identifying measures to prevent such an accident from recurring.

At the end of the analysis, the accidental mechanism is represented in a graphic form, by a tree linking the facts identified as causal elements that led to the occurrence of the accident.

As an example, a tree of events has been built in Part 2, at the end of case study 16: it enables us to visualize the mechanism of an explosion which occurred in a milk spray dryer.

PART 2

27 Case Studies of Domestic or Industrial Explosions

Introduction to Part 2

This second part of the book successively presents the 27 case studies of domestic or industrial explosions that have been investigated.

For each of them, the presentation uses the same three-part plan:

– a description of the installation, the circumstances of the explosion and the damage it caused;

– an analysis of the various elements necessary for establishing the mechanism of the explosion: its circumstances are usually sufficient to identify the nature of the explosive system involved and its quantity can be estimated from the damage analysis;

– a statement of lessons learned from the investigation.

All 27 cases have been chosen in order to be representative of one or another of the four explosive systems presented in Chapter 1, and to cover the widest spectrum of the possible mechanisms.

Table P2.1 gives the different characteristics of each of them:

– the type of explosive system involved, the nature of the reagent(s) involved and their state (gaseous, liquid, solid or dust);

– the type of environment, domestic or industrial;

– the type of industry where the industrial explosions occurred;

– the number of cases for each type of explosive system.

Type of environment (domestic or industrial) and type of industry		State of the product(s) involved	Nature of the product(s) involved	Special feature	Number of the case study	
Type 1 (ATEX explosion)						
Domestic environment		Gas	Natural Gas (NG) or butane	Discrimination between both gases	1	6 cases
			NG	Fire followed by 1 explosion	2	
			Aerosol propellant	Fire followed by 2 explosions	3	
			NG	Mechanism based on leakage rate	4	
			Propane	Mechanism based on leakage rate	5	
			Propane	Mechanism based on location of ATEX	6	
Industrial environment	Paper industry	Gas	Hydrogen	Explosion in a pulp paper tank	7	11 cases
	Energy industry	Gas	Hydrogen	Explosion in an electrolyzer	8	
	Automotive industry	Gas	Propane		9	
	Petroleum industry	Gas	Hydrocarbons (C3 – C4)	Explosion in a refinery	10	
	Metallurgic industry	Gas	Cupola gases		11	
	Pharmaceutical industry	Vapor	Acetone	Leakage of hot liquid acetone outside	12	
	Pharmaceutical industry	Vapor	Toluene	Ignition of ATEX by an electrostatic source	13	
	Cement industry	Vapor	Kerosene		14	
	Automotive industry	Vapor	Hydrocarbons (C5 + heavier)		15	
	Agro-food industry	Dust	Powdered milk		16	
	Wood industry	Dust	Wood dust		17	

Type 2 (reaction between 2 or more reagents or decomposition of a single reagent						
Industrial environment	Chlorine industry	Gas + Gas	$H_2 + Cl_2$		18	7 cases
	Aerospace industry	Gas + solid	$O_2 + Fe$		19	
	Metallurgic industry	Liquid + liquid	Liquid aluminum + water		20	
	Chemical industry	Liquid + liquid	HNO_3 + glyoxal	Runaway reaction	21	
	Chemical industry	Liquid + solid	Molten alkaline nitrates + Fe		22	
	Sugar industry Poultry industry	Liquid + solid	$KMnO_4$ + alcohol or aldehyde	The reaction initiates an ATEX explosion	23	
	Aerospace industry	Vapor	Substituted hydrazines	Explosive decomposition initiated by a gas	24	
Type 3 (hollow Body Burst)						
Industrial environment	Chlorine industry	Liquid + liquid	Cl_2 + another reagent		25	1 case
Type 4 (vaporization of an overheated liquid)						
Industrial environment	Waste industry	Liquid + solid	Water + slag		26	2 cases
	Waste industry	Liquid	Water in a boiler		27	

Table P2.1. *Characteristics of the 27 cases of explosion investigated*

P2.1. Domestic explosions

Explosions that occur in a domestic environment are always explosions of ATEX (explosive system 1) resulting from the leak of a flammable gas, in the more or less confined space of a room (kitchen, boiler room, basement, etc.).

To investigate these explosions, the identification of their causes must deal with the many aspects relative to the numerous parameters, in which the volume of the ATEX formed and the effects of its explosion depend on:

– the leak possibilities of the various systems containing or supplying pressurized gas (LPG cylinders – butane or propane, natural gas (NG) or LPG supply network, etc.) to various appliances (gas cooker, water heater, boiler, gas heater, etc.);

– the section of the leak orifice;

– the existence of devices for limiting and regulating the gas pressure upstream of the leak orifice (length of the leaking line, regulator, burner, torch, boiler or water heater pilot light, etc.) which, in addition to the leak orifice itself, are as many devices to limit its flow rate;

– the gas pressure upstream of the leak orifice;

– the flow rate of the discharged gas;

– the leak rate estimation;

– the estimated volume of discharged gas which, from the estimation of the duration of the leak, generally accessible from the circumstances of the explosion, and from the leak rate, enables to estimate the volume of discharged gas;

– the location of the leak orifice (distance from floor or room walls);

– the environment of the leak orifice and the confinement level of the leaking space;

– the comparative densities of the discharged air and gas;

– the leak rate compared to the room air change rate, whether natural or forced;

– the volume of the ATEX formed;

– the consistency between the volume of the ATEX formed, the mechanical damage caused by the explosion on the building elements of the room (exterior walls, partitions, doors, windows, etc.) and the estimation of the forces exerted by the overpressure resulting from the explosion.

P2.1.1. *Determination of ATEX location and volume*

Two important parameters for determining the location and the volume of ATEX are the flow rate and the density of the gas.

Harris (1983) has studied how a gas leaking out of a leak orifice is distributed in the quiet atmosphere of an unventilated room, depending on whether the gas is more or less dense than air (typically propane and NG). For each of these situations, two cases must be considered according to the conditions of ejection velocity at the outlet of the discharged gas, i.e. according to whether the flow is turbulent or laminar.

The results obtained are as follows:

– whether the speed is laminar or turbulent, the concentration gradient established in the room depends on both the height h_0 of the leakage orifice relative to the ground and the difference in density between the air and the gas;

– the higher the concentration of NG (of propane respectively) at one point in the space, the closer the point is to the ceiling (to the ground, respectively);

– the gas concentration is equal at each point of the same horizontal plane;

– what distinguishes a laminar regime from a turbulent regime is that, in the first case, the flow of gas is slow and the gradient of concentration is under the sole influence of the difference in density of the two gases. In the second case, on the contrary, the turbulence produced by the jet of gas favors the entrainment of air at the level of the leak orifice and thus also the mixture of these two gases, up to far distances from it.

P2.1.1.1. *Case of a laminar flow*

In the case of a propane leak occurring in laminar conditions, Figure P2.1 shows the profile of curve $c = f(h)$ on a vertical, for a leak orifice located 1 m from the ground and for a given time after the leak started:

– $c = 0$ at any point above h_0;

– when h decreases from h_0 to 0, c increases progressively until it reaches, at ground level, a value that is all the higher as the leak lasts longer (in the case of Figure P2.1, the duration of the leak is not specified);

– the ATEX formed is contained in a horizontal slice, the thickness of which is defined by the LEL and UEL values of propane (respectively, 2.1 and 9.5%

volume): it is at an average height between the leak orifice and the ground, and its volume represents only a small proportion of the room volume, given that the range between the LEL and UEL values is small;

– in the case of an NG leak in laminar conditions, c should be expected to increase progressively between the leak orifice and the ceiling, symmetrically with a profile similar to that of Figure P2.1.

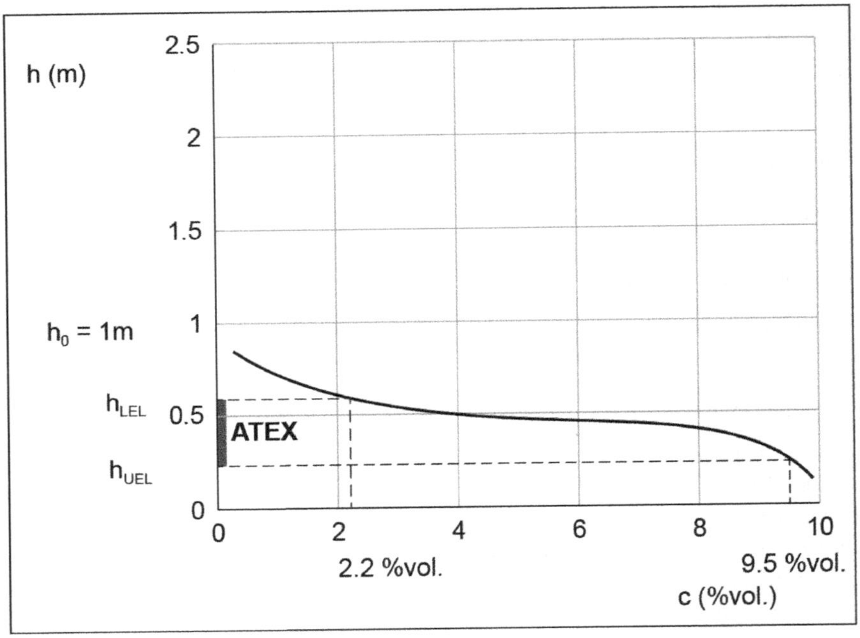

Figure P2.1. *Variation of the propane concentration c as a function of the height h relative to the ground, at a given moment after the beginning of the leak under laminar flow conditions. For a color version of this figure, see http://www.iste.co.uk/chaineaux/mechanisms.zip*

P2.1.1.2. *Case of a turbulent flow*

In the case of a turbulent flow of NG, for a leak orifice located 1.9 m from the ground, Figure P2.2 shows the profile of curves c = f(h) on a vertical, for 20, 40, 60 and 80 mn leak durations:

– Compared to Figure P2.1, it is the turbulence produced at the leak orifice that changes the profile of curves c = f(h):

- whatever the leak duration, it is for a value of h_1 well below h_0 that $c = 0$;

- c increases approximately linearly between h_1 and h_0, up to a value that depends on the leak duration.

– For a given leak duration, c remains constant between the leak orifice and the ceiling, and its value increases with the leak duration:

- c remains below LEL for durations of 20 mn or less, c is between LEL and UEL for durations between 20 and 60 mn, and c is above UEL for durations of 60 mn or more;

- it is therefore the leak duration that determines the presence, location and volume of an ATEX in the room: it appears that no ATEX is formed in the room as long as the leak duration remains less than 20 mn, while an ATEX is present at any point between the leak orifice and the ceiling, for any duration between 20 and 60 mn. Beyond 80 mn, c becomes greater than the UEL at any point between the leak orifice and the ceiling, so that no ATEX is present there, but there is an ATEX in a thin slice under the level of the leak orifice;

- in Figure P2.2, it appears that an ATEX is present over a range of values of h larger for a leak duration of 60 mn (range shown in blue) than for a leak duration of 80 mn (range shown in orange);

- note that the value of c at the ceiling depends not only on the leak rate but also on the room volume; in Figure P2.2, the room volume is not specified and this is the reason why the values given for the leak duration are purely indicative;

- moreover, the absence of ATEX between the leak orifice and the ceiling, for a sufficiently long duration, presupposes that the room atmosphere is calm, with no movement of air other than that resulting from the leak itself. Actually, it would be only if there is no opening in the room (no door, no window or door and window seals that are particularly tight) that the atmosphere would be totally quiet, but this situation never occurs. On the contrary, if the room is ventilated, mechanically or even naturally, it prevents the regular increase of c with time;

- it follows that the volume v of ATEX can only be estimated in an approximate way;

- in the case of a turbulent propane leak, it is expected that for a given leak duration, the propane concentration c will increase between the leak orifice and the ground, symmetrically and with a profile similar to that of Figure P2.2.

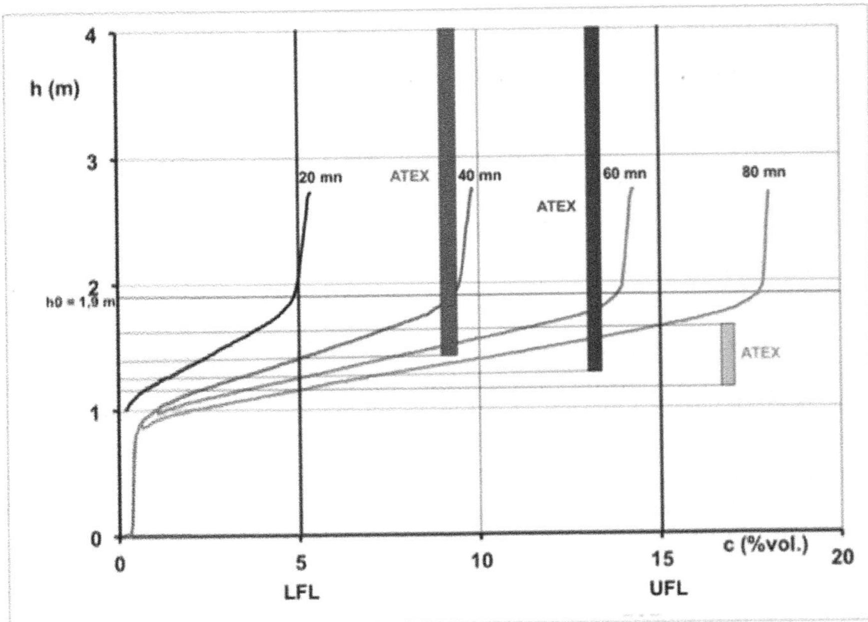

Figure P2.2. *Variation of the concentration c in GN as a function of the height h above the ground, for different values of the leak duration, under turbulent flow conditions and for h_0 = 1.9 m. For a color version of this figure, see http://www.iste.co.uk/chaineaux/mechanisms.zip*

P2.1.2. *Thermal effects of an explosion on buildings and the human body*

As indicated above, the flame of an explosion contains very hot combustion gases, typically more than 1,000 K. Such a temperature is sufficient for a flame propagating in deflagration regime to initiate the combustion of combustible materials present in contact with them or even simply subjected to their radiation for the most flammable materials. Conversely, although containing flue gases a little warmer than for a deflagration, the flame of a detonation is too short to initiate the combustion of these same materials.

Note here that the distinction between deflagration and detonation is rather academic, in that all ATEX explosions in domestic environments are deflagrations:

– in the presence of air, the NG is a gas that, when mixed with air, forms an ATEX where the flame propagates too slowly to detonate;

– similarly, although propane and butane are more reactive gases than NG, it can be argued that an ATEX containing these gases cannot detonate, under the conditions of confinement such as a residential space (this characteristic would be different if the geometry of the confinement were very elongated).

For the thermal effects of an ATEX explosion on people in contact with or exposed to the radiation of combustion gases, the experience shows that any exposed bare skin surface is severely burned and that these people are then seriously or fatally burned, depending on the size of the bare skin surface exposed to the flame.

Conversely, the fact that a person subjected to the effects of an explosion was not burned is an indication that they were not in contact with the combustion gases or under the effect of their radiation, which gives information on the maximum volume of the flame involved. However, it must be taken into account that the flue gases undergo a volume expansion of a maximum factor equal to 10, which corresponds to a radial expansion (ratio of the maximum radius of the flame to the radius of the initial ATEX) with a maximum value slightly higher than 2.

P2.1.3. *Mechanical effects of an explosion in a housing*

In order to assess the consistency between the volume of an ATEX and the mechanical effects resulting from its explosion in the confined space of a housing, the following parameters must be taken into account:

– the maximum overpressure likely to be developed in an explosion-resistant containment of the ATEX which is contained in its volume, this maximum overpressure itself depending on:

- the composition of ATEX,

- the fraction of the containment volume occupied by ATEX;

– the rate of pressure rise of the explosion, propagating in the deflagration or detonation regime (see sections 1.2.3.1 and 1.2.3.2);

– the overpressure for which the most fragile of the containment components would break;

– the volume of ATEX whose explosion can develop enough overpressure to break the most fragile of these components.

It has been noted in section 1.1.4.3.1 that the maximum pressure P_{ex} (the so-called “maximum explosion pressure”) which is likely to be developed during the deflagration of a homogeneous and stoichiometric ATEX, contained under atmospheric pressure in a resistant containment, is slightly less than 10 absolute bars (see Figure 1.5). In addition, the explosion pressure decreases when the NG content moves away from the stoichiometry, both towards the LEL and the UEL.

P2.1.3.1. *Forces exerted on the walls of a room during an ATEX explosion*

A building of conventional structure, constructed of cinderblock or brick, is characterized by a rupture pressure P_r very much lower than 10 absolute bars. A pressure of 10 bar (100 mbar, respectively) corresponds to the exercise of a 100 daN force (1 daN, respectively) exerted uniformly over 1 m^2.

Compared to cinderblock or brick walls, where the weak points are the mortar joints, reinforced concrete walls are made of a more resistant homogeneous material.

However, a parallelepiped room, whose walls are made of reinforced concrete with one reinforcement steel sheet, is not strong enough to withstand the forces developed by an explosion. For example, in a cubic building with a surface area of 10 m^2, i.e. having a 3.3 m edge, an overpressure of 100 mbar corresponds to a force of 1 daN on a surface of 1 m^2, i.e. 10 daN on the total surface of each wall of 10 m^2.

Assuming that each wall of a room is rigid and non-deformable, this force is transmitted entirely in the form of tensile forces exerted on the connections of the wall with the other elements of the structure: for example, connections of a vertical wall with each of the two adjacent vertical walls, with its soil and its ceiling. These are considerable efforts that these links cannot withstand and that lead to ruptures in the plastic field.

To be able to withstand a pressure of only 10 daN/m^2, i.e. 1 bar, a construction must be made of reinforced concrete with at least two reinforcement sheets completed by gussets, to reinforce the connecting dihedrons between the walls, the ground and the ceiling. This type of construction, which corresponds to a kind of casemate, is built in establishments where explosive substances are used, but is obviously not of the conventional type for a housing.

In Chapter 1, it has been considered that the explosion of a stoichiometric homogeneous ATEX, contained in a sphere resistant to the pressure developed by the explosion. If the ignition of the ATEX is initiated at the center of the sphere at ambient temperature and pressure, the maximum pressure P_{max} value reaches a little less than 10 bar (Figure 1.5 shows, for example, the curve P(t) for an air–CH_4 mixture at 9.5% volume CH_4).

It must be added that, at the limits of the explosive range, close to LEL or UEL, the values of the maximum pressure P_{ex} are less than P_{max} but are still worth several bars (see the section 1.1.4.3.1 on the combustion of butane).

P2.1.3.2. *Influence of room openings*

It should also be noted that a housing is never completely free of doors and windows. These elements and their frames are even less resistant than the connections between the walls. These openings are all surfaces through which the flue gases produced by the explosion can be evacuated, thus limiting the overpressure developed in the room and being eventually able to protect it from the destruction of its walls However, the limitation effect of these openings remains negligible, as soon as the ATEX volume represents at least 10% of the room volume.

The very low value of this proportion means that, when an ATEX exploded inside a housing and this building suffered only relatively moderate destruction (broken windows, windows and doors torn), the intensity of this destruction is generally not a suitable element to provide accurate information on the volume and location of the ATEX involved.

On the contrary, if the destruction is much greater (exterior walls destroyed, debris from these walls projected around, ceiling and floor slabs collapsed or cracked), the volume of ATEX involved certainly represents a significant fraction of the volume of the room.

Case 1

Discrimination Between NG and Butane

The objective of the investigation was to determine whether NG or butane was the gas constituting the exploding ATEX, according to these two possible scenarios:

– NG would have leaked from a damaged pipe of the main network, buried under the sidewalk along the front of the building, and this leak would have allowed the gas to migrate through the ground into the basement, to form the exploding ATEX;

– butane would have been discharged from a 13 kg cylinder, stored on the ground floor of the building and would have flowed to the basement where it would have mixed with the air to form the exploding ATEX.

C1.1. Different arguments tentatively used for discrimination

A previous investigation had led us to consider that the effects of the explosion were the characteristic of a detonation.

C1.2. Thermal effects of the flame

The thermal behavior of certain combustible materials (thin paper, thermoplastic film, etc.) that may ignite when they are exposed to the flame of a deflagration had not been observed; on the contrary, it is commonly accepted that these materials do not undergo the thermal effects of the too short flame of detonation. Therefore, since butane is a gas able to detonate, unlike NG, the hypothesis of a butane detonation had therefore been favored. However, this argument does not seem to be justified, because a detonation would certainly have been certainly preceded by a deflagration phase and a transition from deflagration to detonation would have needed a sharp acceleration of the flame but, in the circumstances of the explosion, the geometric conditions for such an acceleration were not met.

C1.3. Mechanical effects of the explosion

The mechanical effects resulting from an explosion are greater for a detonation than for a deflagration, because stronger forces are exerted, as a result of the pressure discontinuity that characterizes the shockwave associated with a detonation. The previous investigation considered that the damage caused by the explosion was very important and could have only resulted from a detonation. In reality, even a deflagration commonly develops overpressures (tens to hundreds millibars) sufficient to destroy or severely damage buildings and, in the present case, the damage observed could not discriminate butane from NG.

C1.4. Relevant arguments used for the elimination of a butane leak

The previous investigation considered a butane leak through the valve of the cylinder, which had been incompletely closed:

– a priori, the leak flow rate depends only on the cross-section of the valve; but, in accordance with the regulations, the cylinder valve is normally equipped with a steel ball which, under the effect of the butane pressure, partially closes the valve, as soon as the butane flow becomes too high, i.e. when the cross-section of the valve reaches a threshold well below the total opening;

– an experimental determination has shown that 1 hour is the duration necessary for a complete discharge of the cylinder, through a full open valve not equipped with a ball valve; on the contrary, it takes 200 h to empty a cylinder equipped with a ball valve, and this duration corresponds to a maximum leak flow rate equal to about 65 g/h;

– the mechanical effects can be explained by the explosion of about 50 m^3 of a air-butane ATEX that contains about 2 kg of butane if it is supposed to have an LEL composition;

– for the maximum flow rate of butane, i.e. 65 g/h, the duration needed to form such a volume is higher than 30 h, since a significant part of butane discharged is mixed with air under the LEL;

– this very long duration means that a sufficient volume of ATEX would never have been formed because, during the discharge of butane, it would be simultaneously diluted by the air of the basement.

C1.5. Identified mechanism

Finally, the explosion has not been the consequence of a butane leak discharged from a cylinder, but it was the consequence of an NG leak from a pipe of the public network, buried under the sidewalk along the building and of a migration of the NG to its basement.

Case 2

Determination of the Mechanism of an Accident Involving a Fire and an Explosion

It had to be discriminated between an explosion and a fire as the initial phenomenon of an accident which occurred in the basement of a residential building.

C2.1. Circumstances and effects of the explosion

An explosion occurred in a residential building, when a resident came out of his flat and switched on the common parts of the ground floor.

The analysis of damage showed that the thermal effects were very significant in the basement since different things have been marked by the flame: the wooden parts of a piece of furniture and of the staircase, as well as clothing burnt and even a lead pipe melted. Such damage has not been able to be produced by the flame of the explosion during its propagation and, in addition to the explosion, a fire also occurred. In particular, the issue of investigation was whether the explosion initiated the fire or whether the fire caused a leak of gas from the melted pipe supplying the NG boiler (there was no other product that could explode in the basement), and this is why the fire started.

C2.2. Occurrence of a fire prior to the explosion

The mechanism of the accident has been the following one:

– firstly, a fire occurred in the basement (the source of ignition was not identified), and it was intense enough to cause the lead pipe to melt;

– this melting created a leak orifice, through which an ignited jet of NG leaked into the basement, and the fire propagated to the combustible materials present;

– the fire was initially fed by the air of the basement thus the question arises as to whether the air reached the basement from the ground floor and continued to feed the fire;

– if this had been the case, the jet of NG would have remained ignited, no ATEX would have been formed and no explosion would have occurred;

– on the contrary, it can be supposed that, if no fresh air had reached the basement, the oxygen would have become scarcer, the fire would have decreased in intensity and ended up extinguishing itself by default of oxygen;

– the fumes, i.e. the gases produced by the combustion of hydrocarbon materials in highly under-oxygenated conditions, were containing other flammable gases such as CO and H_2;

– because the mixture of these fumes and the NG which were continuing to leak (being very hot and less dense than air), had gone up from the basement, mixed with air from the ground floor and formed an ATEX;

– an analysis of the important damage produced to the building indicates that the ATEX likely occupied a large part of the ground floor, and possibly even reached the first floor.

Case 3

Determination of the Mechanism of an Accident Involving a Fire and Two Explosions

An accident involving successively a fire and two explosions had to be investigated. The first explosion occurred in the cellar of a house where some work had taken place in the morning and interrupted for a lunch break. It had been immediately followed by a fire and later by a second explosion. Among the elements of cause necessary to determine how these three phenomena have succeeded each other, the nature of the combustible products involved in each of them had to be identified.

C3.1. Nature of the flammable gases or liquids involved in the first explosion

The following flammable gases and liquids were or could have been present in the cellar at the time of the first explosion:

– there were a fuel oil boiler and a tank containing fuel oil, but the flash point of this liquid is much higher than the ambient temperature and an explosion of its vapors is not possible in ambient conditions;

– a buried NG line passed along the house; it could have leaked and the NG could have diffused through the ground up to the cellar, but such a phenomenon was not possible in that case and this hypothesis has thus been excluded;

– the propane contained in a cylinder had been used in the morning for welding operations and could have been released during the lunch break, but the absence of the cylinder in the cellar at this time has been proved;

– finally, there were also several aerosol cans containing flammable propellants; one of these could have been damaged (they have been found to have burst after the accident) and have released a very volatile flammable liquid phase, without the knowledge of the workers absent during the lunch break (a leakage from a corroded aerosol could also be possible), which would have evaporated.

It is this hypothesis that has been retained, especially since an estimated quantity between only 20 and 100 g of this liquid phase, mainly consisting of propane-like flammable gases, would have been sufficient to form the exploding ATEX.

The ignition source of the ATEX has been certainly a spark that appeared during the normal operation of the electric starter relay of the boiler, especially since the inhabitants testified that the first explosion occurred at the very moment of the re-ignition of the boiler.

C3.2. Determination of the explosion mechanism

The different circumstances of the accident led to establish that the accident took place according to the following mechanism:

– it was the flame of the explosion that ignited the few combustible materials in the cellar, including the liquefied gases of the aerosol can being evaporated, but not the boiler fuel oil;

– initially, oxygen present in the air of cellar allowed the development of the fire, but it became scarce since the cellar was completely devoid of ventilation, and the fire diminished in intensity until the firefighters arrived. They only had to put out a very small fire;

– when a fire develops in a very under-oxygenated environment, it produces fumes that are composed of soot as well as flammable gases such as CO, CH_4 and H_2; in this case, these fumes had gone up to the ground floor, where they mixed with the air to form an ATEX that exploded in the kitchen and in the veranda (such a phenomenon is well known by firefighters who call it a "backdraft");

– an analysis of the mechanical and thermal damage produced by each explosion showed that the second explosion was significantly more violent than the first one:

- the flame of the first explosion did not come out of the cellar, as the person next to the staircase was not burned,

- the overpressure produced by the first explosion broke the windows of the kitchen and the veranda, but it was not intense enough to crack the walls of the house up to the second floor and produce an air pressure wave intense enough to hurt the two people in the kitchen,

- it is the second explosion that produced the cracking of the walls and that also injured the eardrums of one firefighter present outside near the veranda; the second explosion was more violent than the first because it involved an ATEX present on the ground floor and containing CO, CH_4 and particularly H_2. For this gas, the flame propagates much faster than that of propane (its fundamental burning velocity S_u is worth 312 cm/s, while methane, propane and CO ones are between 40 and 46 cm/s), since H_2 explodes much more violently than these gases.

Case 4

Determination of the Mechanism of an Explosion from the Leak Flow Rate of NG

The investigation consisted of determining the origin of the leak that allowed the formation of an air-NG ATEX in the flat of the second floor of a residential block. To determine the mechanism of explosion, the very large-scale damage produced as well as the circumstances of the formation of a big volume of ATEX have been considered:

– the explosion resulted in three fatalities and one serious injury (severe burns), with the following destruction:

- on the second floor, the flat where the leak occurred and the contiguous one had their ceiling and floor slab collapsed and their exterior walls were partly blown out;

- the ceiling of the first floor collapsed too;

- in addition, debris from the exterior walls was thrown up to several tens of meters, around the building.

The resident of the flat where the gas leak occurred testified that the leak involved the NG supply to his gas cooker installed in the kitchen, and asserted successively that the process of the explosion corresponded firstly to one and secondly to the other of the following two scenarios:

– He voluntarily disconnected the hose fitted to the downstream of the shut-off valve, which he kept in the open position (see the scheme of Figure C4.1). Upstream of this valve, the rigid pipe supplying NG to the cooker was equipped with a pressure regulator (see Figure C4.2).

For a color version of all figures, in this case study, see www.iste.co.uk/chaineaux/mechanisms.zip.

– Later on, he asserted that the leak was due to another scenario: he had inadvertently extinguished two burners of the cooker, letting the gas leaking into the kitchen through their gas valves kept open.

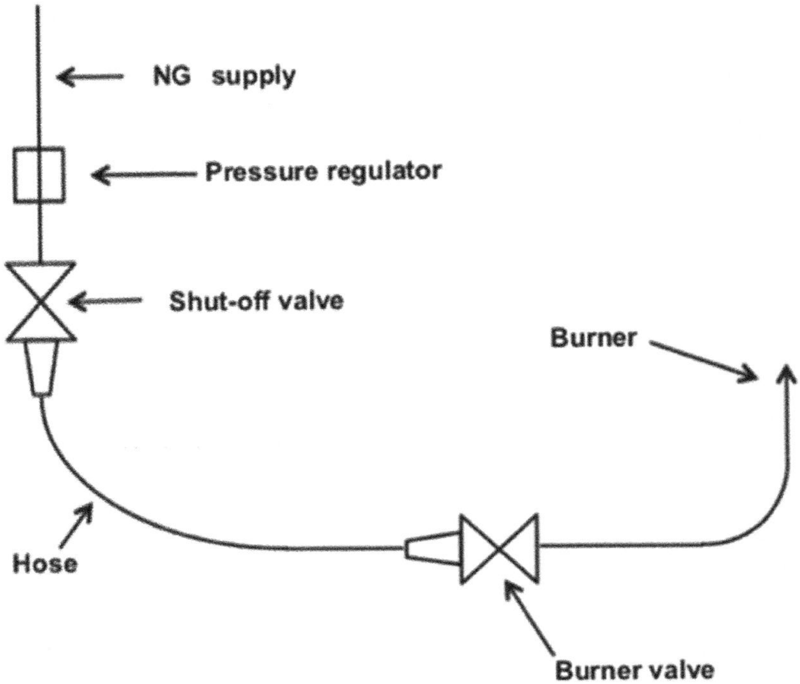

Figure C4.1. *Scheme of the NG supply*

– The diameters of each leak orifice, as well as the maximum values of the gas flows compatible with the pressure regulators located upstream of these orifices, allowed us to estimate the leak flow rates for each scenario:

- in the event of a leak at the end of the disconnected hose, as a vertical jet directed from up to down, the flow was limited by that of the regulator installed at about 1 m upstream of the shut-off valve, i.e. approximately 6 m^3/h for a pressure of 21 mbar;

- in the event of a leak at two burners of the cooker, the leak flow rate could not exceed approximately 0.4 m^3/h, which is the sum of the maximum flows allowed by each burner (see Figure C4.3, which shows an example of a gas burner and injector for an NG cooker and see Figure C4.4, which shows a pressure regulator supplying a gas appliance from a butane cylinder).

Figure C4.2. *Low pressure regulator for the supply of NG to a gas apparatus, up to 67 kW of thermal power (maximum flow rate 6 m3/h, upstream pressure 0.5–4 bar, expansion pressure 21 mbar)*

Figure C4.3. *Gas burner and injector for NG (medium fire, 200 L/h, 2 kW)*

According to the circumstances of explosion and the testimonies, the leak lasted 2 to 5 hours and, according to the leak flow rates for each scenario, the volumes of

NG discharged were worth, respectively, 2 and 30 m^3 at most (corresponding to ATEX volumes 20 and 300 m^3 at most, for a stoichiometric composition).

Figure C4.4. *Pressure regulator supplying a gas appliance under 27 mbar (gas cooker, water heater, etc.) from a butane cylinder*

Although the characteristics of the leak could not be known with certainty, the location of the damage caused by explosion throughout the second flat indicates that the ATEX formed occupied a large volume, which was not limited to a part of the kitchen, but which extended in the living room (the inhabitant who was there at the time of explosion was severely burned) and probably also in the other rooms of the flat (vestibule, bedroom, etc.) or even partly in the common parts of the second and third floors.

The ATEX formed could only result from a 6-m^3/h leak flow rate for several hours, following the disconnection of the gas supply line and not from the flow rate that normally feeds two burners.

It must be added that the magnitude of the damage produced does not result from an abnormal weakness of the building, particularly the mechanical connections between the load-bearing walls and the ceiling of the second floor, but it only results from the weak resistance of a conventional building to the forces exerted by an explosion, as soon as the volume of ATEX represents a sufficient percentage of the total volume of the rooms where the ATEX is present.

Case 5

Determination of the Mechanism of a Propane Explosion from the Leak Flow Rate

An explosion occurred during the intervention of a maintenance agent on a propane boiler, installed in the boiler room of the basement of a two-story building. The objective of the investigation was to determine which of the following two causes the explosion resulted from:

– a malfunction in the design or operation of the facility that supplied the boiler from a tank installed outside the building;

– an error made by the agent during his intervention.

The explosion occurred during the restarting of the boiler, after the agent had performed the standard exchange of the low-pressure regulator, mounted just upstream of the boiler burner.

Because of this exchange, air had entered the propane supply line and the agent had to purge it, prior to restarting the boiler:

– to perform this purge, the agent had to let propane draining the line for some time and to discharge it in the boiler room;

– he could have chosen to disconnect the line just downstream of the low pressure regulator, to limit the purge flow to the maximum allowed by this regulator, approximately 2 g/s;

– but, probably to purge the line faster, he preferred to disconnect it just upstream of the low pressure regulator. This configuration allowed him to drain the line 15 times faster, at the maximum flow rate allowed by the medium pressure regulator located upstream of the lower one, approximately 30 g/s;

– it was during this purge that the explosion occurred;

– from the extent of the damage to the building, it is possible to estimate that the ATEX formed had a volume at least 20 m^3, which leads to estimate between 1 and 2 kg the mass of propane discharged and 1 minute or less the purging time required to discharge such a mass, at a rate of 30 g/s;

– therefore, it appears that the agent significantly underestimated the flow rate of propane during a purge under these conditions since, to operate safely, he would have had to limit the purge time to only a few seconds.

In this case, it is a comparison of the leak flow rate corresponding to each configuration that enabled to determine the mechanism of the explosion.

Case 6

Determination of the Explosion Mechanism, Based on the Location of the Ignition Source of ATEX

The mechanism of an explosion that caused several victims and severely damaged housing had to be investigated.

As it has been pointed out before (see section. 2.1.3.2), an analysis of the nature and intensity of damage does not generally provide accurate information on the volume and location of this ATEX, particularly if the damage is not very significant (broken windows and doors, without destruction of external walls nor floor or ceiling slabs).

Moreover, the mechanism of the explosion is easier to establish if an analysis of the explosion circumstances provides information from which can be deduced the location and volume of the ATEX, in the room where it was formed.

C6.1. Circumstances of the explosion

In the present case, the ATEX could only have been formed as a result of a propane leak in the kitchen, from one of the two following appliances:

– a wall-mounted boiler;

– a gas cooker.

The day before the explosion, the boiler had been serviced by a maintenance agent and after the explosion an anomaly had been observed at the gas inlet valve to the burner which led to the hypothesis that this maintenance operation could have been the cause of a leak of the boiler, at an estimated flow rate of 800 dm^3/h.

However, because another anomaly had been observed with the oven of the gas cooker the valve of which had been found open after the explosion, an assumption was made that the origin of the explosion was a leak in the burner of the oven, at a flow rate estimated at 130 dm^3/h, i.e. about six times lower than that of a boiler leak.

More information on the explosion circumstances were provided by the resident:

– he went to bed at about 11 pm without noticing any anomaly, but he could smell gas the next morning, as soon as he entered the living room, where he said the communicating door to the kitchen was open;

– when entering the kitchen, he saw that the pilot burner was extinguished;

– the explosion occurred while he was about to light one of the cooker burners with a match, and this indicates the ATEX was present approximately between 1 and 1.3 m above the ground;

– the resident had a naked torso and the flame of the explosion burned him severely.

C6.2. Discrimination between the boiler leak and the cooker oven leak

To discriminate between the possible leaks in the boiler or in the oven of the cooker, the characteristics of each leak, i.e. its aerodynamics, its flow rate and its location, had to be taken into account:

– The duration of the leak remains unknown, but in no case could it have exceeded 10 h and, based on the leak flow rates of the boiler and the cooker oven, the maximum amounts of propane discharged would have been 8 and 1.3 m^3, respectively.

– For such values, including the smallest one, it is likely that the ATEX was not only present in the kitchen but also in the living room; it was formed as a result of a competition between the following two phenomena:

1) *The mixing of air with propane coming from the leak orifice*: the air mixed with propane discharged from the leak orifice, in a flowing regime depending on its environment: for the oven burner where propane was discharged into the confined space of the oven, as for the boiler burner where propane was discharged into the combustion chamber, it can be considered that the flowing regimes of the

propane–air mixture coming out of one of these confined places into ambient air could have been similar and close to laminar conditions.

2) *The dilution of the atmosphere of the kitchen with the air coming from outside*: the atmosphere of the kitchen was diluted with the air coming from outside and naturally diffusing through the vents present in the walls (at least one vent is mandatory for a kitchen where there is a gas apparatus) or around the kitchen doorframes. Based on the kitchen volume of approximately 15 m^3 and the kitchen air exchange rate of approximately 1 h^{-1}, the flow rate of fresh air into the kitchen was approximately worth 15 m^3/h.

By referring to the work by Harris (1983), the variation of the concentration of propane as a function of the height above ground of a leak orifice would have the profile of the curve in Figure P2.1, for a laminar regime: in particular, the ATEX would be present between the leak orifice and the ground but not above the leak orifice.

The locations above ground of the match, of the leak orifice of the boiler and of the leak orifice of the oven can be estimated as follows:

– the match was approximately between 1 and 1.3 m above ground;

– the leak orifice of the boiler was approximately between 1.2 and 1.4 m above ground;

– the leak orifice of the oven was approximately between 0.5 and 0.8 m above ground.

Since the ATEX was ignited by the match, it can be asserted that the explosion resulted from a leak in the boiler and not in the oven.

Moreover, since the mixture of air entering the kitchen and propane leaking necessarily had to be flammable, the above result should be evaluated by a comparison between the estimated values of the different flow rates:

– 800 dm^3/h for the leak in the boiler;

– 130 dm^3/h for the leak in the oven;

– 15 m^3/h for the fresh air entering the kitchen.

The LEL of a propane–air mixture is worth 2.2% volume, and for such a mixture the ratio of propane concentration to air concentration is worth 2.2/97.8 = 0.02. This value is comprised between the ratio of the leak in the boiler to the flow rate of fresh air entering the kitchen, i.e. 0.053, and the ratio of the leak in the oven to the flow rate of fresh air entering the kitchen, i.e. 0.009. This result means that the leak in the boiler would actually have given a flammable mixture, unlike the leak in the oven, and it validates the hypothesis of an explosion due to the boiler leak instead of the oven leak.

Lessons learned from the investigation of domestic explosions

An explosion in a domestic environment is always the result of a leak of a gas (NG, propane or butane) used as the fuel in a combustion apparatus (boiler, water heater, gas cooker, etc).

When it is established that, in an accident, an explosion and a fire occurred, one of the issues of investigation is to determine the anteriority of one phenomenon in relation to the other. Such a question is relevant because there have been cases where an explosion has been followed by a fire and also where a fire has been followed by an explosion.

The sequence of the following two phenomena is possible:

– either an explosion initiates a fire: this is a fairly common sequence when combustible products (volatile liquids and low-density solids such as plastic film and paper sheets) are subjected to the thermal effects of the flame of the explosion;

– either a fire initiates an explosion: this is the case when the fire of a hydrocarbon solid develops in the absence of ventilation, and therefore in an increasingly under-oxygenated environment. The fumes produced contain increasing proportions of flammable gases such as CO and H_2. This phenomenon is followed by the following two:

- the absence of oxygen eventually extinguishes the fire,

- the mixing of flammable fumes with air, in favorable proportions, can form an ATEX and the ignition of this ATEX can produce a "backdraft" phenomenon, which is particularly dangerous for firefighters.

The mechanical resistance of a housing to overpressure from an explosion is quite low since, if it has been built with conventional materials, it will suffer significant damage as soon as the overpressure level reaches a few tens of millibars. Such an overpressure value can be reached in case of an explosion where the volume of the exploding ATEX only represents a few percent of the room volume.

It follows that:

– if relatively limited damage (broken glass, broken door frames, cracked or knocked-down bulkheads) has occurred as a result of the explosion, it is generally not possible to accurately estimate the volume and location of the exploding ATEX, nor therefore to deduce the mechanism of formation of the ATEX;

– if significant damage has been caused (destruction of exterior walls and projection of debris several tens of meters, rupture of openings and their frames, rupture or cracking of ceiling and floor slabs, etc.), it is likely that the ATEX volume represented a significant proportion of the total volume of the floor of the housing, or even of the adjoining ones;

– in many cases, the circumstances of an explosion need to be examined very carefully, as they usually provide essential information to determine the mechanism of the explosion;

– this is particularly true when these circumstances relate to the possible duration of the gas leak and its flow rate, according to the devices present on the gas distribution system, upstream of the leak orifice. Of these, a regulator is used to regulate this upstream pressure of a gas apparatus burner, but it must be taken into account that it also conditions the gas flow rate.

Case 7

Explosion of a Hydrogenated ATEX in a Pulp Paper Tank

C7.1. Description of facilities, circumstances and effects of the explosion

In the paper industry, a significant fraction of the production comes from recovery boxes that are processed through successive steps of crushing in hot water at 50°C, purifying the pulp obtained, and then splitting the pulp into short fibers and long fibers. These two types of fibers are treated separately, in order to be thickened and stored, before being sent to paper machines. Long fibers are stored in a 1,000-m^3 cylindrical tank with a conical roof (about 20 m height and 9 m in diameter). Depending on the manufacturing phases, this tank may be in the filling or drawing phase, or even in the simultaneous filling and drawing phases, so its pulp level may vary between certain limits. Moreover, during the tens of hours before explosion, the filling rate of the tank remained close to 80%.

The tank is equipped with a vent located in its roof, which allows the atmosphere of its sky to communicate freely with ambient air and to remain at atmospheric pressure.

An explosion occurred in the sky of this tank, while an operator was cutting-off a pipe, at approximately 1 m from its connection point to the roof. An overpressure occurred in the tank, tearing the welding of the roof over more than half of its circumference (see Figure C7.1), and resulting in the deadly fall of the operator. The explosion could only result from the formation and subsequent ignition of an ATEX formed in the sky of the tank.

Figure C7.1. *View of the torn tank roof. For a color version of this figure, see www.iste.co.uk/chaineaux/mechanisms.zip*

C7.2. Objectives of the investigation

The objective of the investigation is:

– to determine the composition of the ATEX formed and the conditions of this formation;

– to identify the ignition source of the ATEX;

– to examine whether the expected effects of the identified ATEX explosion are consistent with the observed effects.

C7.3. Determination of the composition of the ATEX

C7.3.1. *Experimental determination of the ATEX components*

Laboratory tests were carried out in a 5-L biomethanizer on paper pulp samples, taken from the operating facilities, in order to determine their behavior when they are stirred and heated at 40°C under air, and to identify the ATEX components:

– for a dryness of the pulp close to 5% (i.e. containing 95% of water), a hydrogen release was identified (it could have a bacterial origin), in such a quantity as, after 20 hours, an ATEX is formed in the sky of the biomethanizer, having an hydrogen content equal to the LEL, i.e. 4% volume;

– according to this result and to the volume of the biomethanizer sky, the order of magnitude of the hydrogen release flow into the tank sky was extrapolated to be about 20 L/h per 1 ton of dry pulp;

– a CO_2 release was also identified and, from the levels of H_2 and CO_2 in the biomethanizer sky, the CO_2 flow was estimated to be five to six times more than the H_2 flow, so that the CO_2/H_2 flow ratio was estimated to be 5;

– by extrapolation of the operating conditions, the release flow of the H_2–CO_2 mixture in the tank sky was estimated to be 1.6 m^3/h.

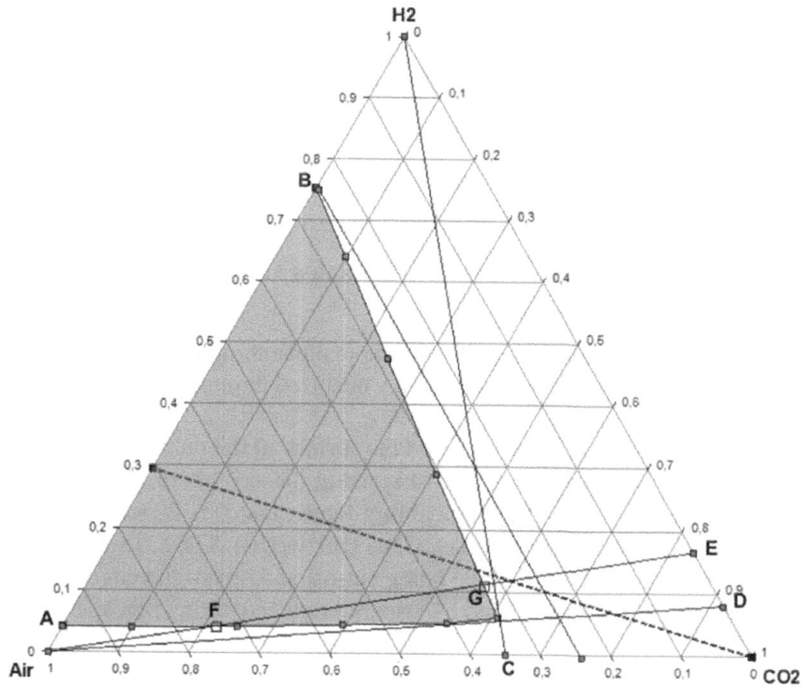

Figure C7.2. *Flammability diagram of H_2–air–CO_2 ternary mixtures*

The remarkable explosive characteristics of H_2–air–CO_2 mixtures can be drawn from Figure C7.2:

– the values of the LEL and UEL of H_2 in air are 4% (point A) and 75% volume (point B);

– the limiting air concentration below which no H_2–air–CO_2 mixture is flammable, regardless of its H_2 content, is 25% volume and the limiting oxygen concentration (LOC) is 5.2% volume;

– the limiting value of the CO_2/H_2 ratio below which no H_2–air–CO_2 mixture is flammable, regardless of its air concentration, is 11.3;

– the LEL and UEL of an H_2–air–CO_2 mixture such as the CO_2/H_2 ratio = 5 are derived directly from Figure C7.2, i.e. 26 and 67% volume, respectively.

C7.3.2. *Determination of the ATEX conditions, formation and ignition*

Before the explosion, a 9-day period elapsed during which the filling rate of the tank, which contained a pulp at about 10% dryness, remained relatively constant and close to 80%, and the volume of sky remained close to 225 m^3. According to the value of the release rate of the H_2–CO_2 mixture in the tank, it took only 36 hours for the sky atmosphere, supposed to be homogeneous, to reach the LEL of the released H_2–CO_2 mixture, i.e. 26% volume. In fact, the atmosphere of the tank was certainly not homogeneous for the following reasons:

– the H_2–CO_2 mixture evolved from the surface of the pulp;

– compared to CO_2, H_2 is characterized by both a very lower density and a very higher diffusion rate;

– throughout the period elapsed before the explosion, an exchange between the outside air and the atmosphere of the tank sky occurred through the vent on its roof.

If it is accepted that there was no air on the surface of the pulp and that there was no H_2 by the vent, there was necessarily a space in the sky where the content of the H_2–CO_2 mixture belonged to the flammability range, and it is therefore possible to say that a fraction of the atmosphere inside the tank was an ATEX. In particular, this ATEX spread to the inside of the pipe being cut-off and, given the circumstances of the explosion, it was certainly ignited by a spark produced during this operation.

C7.3.3. *Consistency between the explosion effects and the estimated ATEX volume*

The tank was made of a 4-mm thick stainless steel sheet, welded to a conical roof of the same material, as a sharp angle welding which was the most fragile zone of the tank and which was torn apart. From the thickness and the diameter of the ferrule, e = 4 mm and D = 9 m, respectively, and from the R_m break stress of stainless steel, i.e. about 400 MPa, Lamé's relationship enables us to calculate the relative rupture pressure $P_r = 2eR_m/D$ which would have led to the rupture of the ferrule, according to a generatrix ("buttonhole" rupture), if the welding had not broken: the calculation leads to a P_r value of around 1.5 bar.

Since the welding broke, this break occurred at a pressure of 1 to 1.5 bar. To check the consistency between the mechanical effects of the explosion and the volume and composition characteristics of the ATEX, it must be shown that the explosion could develop a relative pressure in the tank between 1 and 1.5 bar. However, the calculation of the relative explosion pressure P_{ex} of an ATEX having the LEL concentration, for adiabatic conditions, led to the value P_{ex} = 1.2 bar, and this value would be higher for a H_2–CO_2 mixture having a higher concentration than the LEL.

C7.4. Conclusion

From this analysis, it can be concluded that:

– even though the ATEX was not present in the whole volume of the sky, it was present in probably a large fraction of this volume, because of the large explosive range of H_2–CO_2 mixtures, and it is relevant to attribute the mechanical effects on the tank to the explosion of that ATEX;

– however, it is not possible to precisely estimate its volume or composition.

Lessons learned from the investigation of the explosion of a hydrogenated ATEX

It is necessary to identify all the situations where hydrogen may be in the presence of air, even though this presence is unpredictable or seems improbable, taking into account the processes used and the devices used:

– the exploitation of the following processes can lead to the production of hydrogen, more or less abundant:

- any process where a bacteriological reaction, such as acetogenesis, can occur, such as that observed in the pulp storage tank,

- any process involving highly heated hydrocarbon products (combustion in an under-oxygenated environment or pyrolysis in the absence of air); note that the production of hydrogen is then accompanied by a concomitant production of CO;

– the use of the following devices is accompanied by a production of hydrogen which, under abnormal operating conditions, may be sufficient to form a significant volume of ATEX:

- certain types of batteries during recharging,
- electrolyzers.

Finally, we must also mention the situations where hydrogen is transported by pipe between a storage facility and the facility where it is used. In the transport line, hydrogen is generally under a pressure above atmospheric, so that a leak of hydrogen may occur. An ATEX is formed close to the leak point and its volume depends on the pressure of hydrogen and the section of the leak orifice.

The prevention of an ATEX formation, in any situation similar to that of the pulp tank where hydrogen evolves from a product in an air container, can be achieved by diluting hydrogen with air.

This air must be continuously introduced into the container, at a sufficient flow rate, i.e. close to 100 times the flow of hydrogen. In addition, the air must be introduced as a jet directed to the bottom of the container and sufficiently violent to reach the surface of the product from which the hydrogen is released.

Case 8

Explosion of a Hydrogenated ATEX in an Electrolyzer Cell

C8.1. Description of facilities and explosion circumstances

The function of an electrolyzer is to ensure the decomposition of the water molecule, in each of its components, hydrogen and oxygen, by supplying electrical energy; these two gases are produced at the cathode and the anode, respectively, in the ratio of two volumes of hydrogen per one volume of oxygen. Each cell, as well as some elements of the collection networks and structure, is made of PTFE.

An explosion occurred in several adjacent cells of an industrial electrolyzer, in operation at approximately 130°C under 30 bar, and under normal conditions of use.

C8.2. Effects of the explosion

Several electrolyzer cells were damaged by overpressures having probably resulted from two consecutive explosions of a hydrogen–oxygen mixture ATEX that produced both mechanical and thermal effects. The structure of the electrolyzer was deformed, reflecting the fact that it had been subjected to sufficient pressure to cause the creep of PTFE, a phenomenon that occurs at room temperature when it is subjected to pressure of several tens of bars. With regard to the thermal effects of the explosion, the local melting of PTFE parts was observed.

C8.3. Investigation objectives

The objective of the investigation is:

– to determine the conditions for the ATEX formation and its location;

– to determine the ignition source of the ATEX;

– to establish consistency between the estimated value of the developed overpressure and the observed mechanical effects.

C8.3.1. *Formation and location of an ATEX in the electrolyzer*

For reasons of congestion, it is advisable to construct an electrolyzer as an assembly of a large number of small cells and to collect the gases produced by each cell to two accumulators, using two separate collection networks. Each cell is equipped with a membrane separator, which prevents the mixing of hydrogen with oxygen and the formation of an ATEX, particularly in each of the networks. However, because of the limited efficiency of this device, it is not possible to be sure that the hydrogen (respectively oxygen) flowing in its network is totally free of oxygen (respectively hydrogen). Therefore, if it is supposed that a hydrogen–oxygen mixture flows in each network, it is essential to check that the hydrogen content of this mixture is sufficiently lower than the LEL or higher than the UEL of hydrogen in oxygen, i.e. 4 and 95% volume, respectively.

In the case of the accident, the local degradation of one cell separator explains that an ATEX was formed there, as well as in the hydrogen collection network, downstream of the cell.

C8.3.2. *Results of experimental study*

We carried out an experimental study of the operation of a simplified electrolyzer cell, the anode, cathode and electrolyte of which were in the presence of an atmosphere consisting of a hydrogen–oxygen ATEX; this ATEX was contained under 30 bar, in a 1-L container heated to 130°C, able to withstand a pressure of several hundreds of bars. When the electrodes were powered up, the ATEX was ignited several times by incandescences that appeared on the cathode surface, apparently randomly, at locations where the cathode was not completely wet by the electrolyte.

C8.3.3. *Consistency between the mechanical effects and the overpressure*

The maximum overpressure measured during the tests, i.e. about 60 bars, explains the creep of the PTFE in the industrial electrolyzer. Similarly, the local melting of the PTFE is easily explained by the fact that the flame temperature of a hydrogen–oxygen ATEX exceeds 3,000 K. It should be pointed out that the overpressure developed by the explosion of an ATEX contained in a containment is independent of its volume, when the containment has a compact shape or, if it is elongated, when the combustion flame propagates rapidly, as is the case for a hydrogen–oxygen one.

Lessons learned from Cases 7 and 8

It is necessary to identify all the situations where hydrogen may be in the presence of air, even though this presence is unpredictable or seems improbable, taking into account the processes used and the devices used:

– the exploitation of the following processes can lead to the production of hydrogen, more or less abundant:

- any process where a bacteriological reaction, such as acetogenesis, can occur, such as that observed in the pulp storage tank,

- any process involving highly heated hydrocarbon products (combustion in an under-oxygenated environment or pyrolysis in the absence of air); note that the production of hydrogen is then accompanied by a concomitant production of CO;

– the use of the following devices is accompanied by a production of hydrogen which, under abnormal operating conditions, may be sufficient to form a significant volume of ATEX;

- certain types of batteries during recharging,

- electrolyzers.

Finally, we must also mention the situations where hydrogen is transported by pipe between a storage facility and the facility where it is used. In the transport line, hydrogen is generally under a pressure above atmospheric, so that a leak of hydrogen may occur. An ATEX is formed close to the leak.

The prevention of an ATEX formation, in any situation similar to that of the pulp tank where hydrogen evolves from a product in an air container, can be achieved by diluting hydrogen with air.

This air must be continuously introduced into the container, at a sufficient flow rate, i.e. close to 100 times the flow of hydrogen. In addition, the air must be introduced as a jet directed to the bottom of the container and sufficiently violent to reach the surface of the product from which the hydrogen is released.

Case 9

Explosion of an Air–Propane ATEX

C9.1. Case presentation

The issue of the investigation was to determine the mechanism of an explosion that occurred at a company that was engaged in repairing automotive radiators, by welding and brazing, using propane-powered torches.

The explosion completely destroyed a building housing these facilities and severely damaged buildings in the surrounding industrial area.

C9.1.1. *Description of the facilities*

The building, oriented north–south, 24 m x 18 m for a volume of 2,400 m^3, consisted of walls made of concrete panels, a fibrocement roof, carried by main concrete trusses and by secondary steel trusses.

Several installations were supplied with propane from a 4-m^3 capacity tank, buried outside at the foot of the building wall and equipped with a first regulator that ensured a maximum flow rate of 100 kg/h, under a relative pressure of 2 bar:

– several radiant tubes installed under the roof to heat the building;

– a set of four metal melting stations, equipped with burners;

– a set of five identical welding stations equipped with torches.

Propane was supplied from the tank via a 26/28-mm diameter mainline which, at its point of entry into the building, was divided into a branched network of 12/14-mm diameter lines, serving, in particular, welding stations located in the southwest corner of the building.

Each welding station was equipped with the following two types of torches:

– a "disassembler" torch: this was powered directly by propane, without a regulator, i.e. under 2 bar of relative pressure. Its line was equipped from upstream to downstream with the following:

- a quarter-turn wall valve located at the end of the 12/14 rigid line in front of the welding station,

- fixed by a clamp on a nipple fitted to this valve, a flexible elastomer hose of about 10 mm in diameter, a few meters in length;

– a welding torch, which is used for soldering operations, and is supplied with propane via a wall-mounted device located at the end of the rigid pipe that performs the following three functions:

- shut-off valve function ¼ turn,

- second propane expansion regulator function, from 2 bar to 37 mbar, for a maximum flow rate of 1.3 kg/h. As a safety function, in the event of an abnormally high flow rate, due, for example, to a leak and resulting in a strong pressure drop downstream, the device closed the propane supply,

- downstream of the wall regulator, an elastomer hose identical to that supplying a disassembler torch was attached by means of a hose clamp.

C9.1.2. *Circumstances of the explosion*

The explosion occurred late in the evening of a working day, with no apparent warning signs, approximately 2 h 30 min after the last person left the building.

Three of the five valves that fitted the disassembler torch lines were found in the rubble and the fact that two of them were in the open position can be considered as circumstances of the explosion, since they were necessarily open before the explosion.

C9.1.3. *Explosion damage*

The explosion completely destroyed the roof and walls of the building, leaving only the two pillars that bore its southern façade, as well as all the two pillars of its northern façade and end farm, and finally, some metal trusses on the north side. As for the concrete panels that formed the walls, some entire panels close to the south façade were projected up to about 10 meters, while fragments of variable surface were projected at a greater distance.

In the surrounding industrial area, the explosion severely damaged the buildings closest to the center of the explosion and broke glass up to 400 m away.

After having identified the position of damage on a map of the site, each type of damage has been defined by a couple of values:

– the maximum overpressure of the aerial pressure wave which produced this damage on this building;

– the distance between the building and the center of the explosion.

C9.1.4. *Establishment of the explosion mechanism*

The case discussed here corresponds to that described in section 2.7.2:

– the implementation of propane in the damaged facilities is a circumstance of the explosion that imposes the nature of the explosive system involved, to the exclusion of any other system;

– it is therefore possible to say that the explosion resulted from the ignition of an air–propane ATEX formed in the building as the result of a propane leak.

To establish the mechanism of the explosion, a detailed analysis of the circumstances of the explosion and the effects it produced should be used to clarify certain elements of the mechanism, such as:

– the conditions for the formation and ignition of the ATEX (origin of the leak, nature and location of the ignition source, etc.);

– the estimated mass m of propane contained in the ATEX.

At the end of this analysis, it must then be verified that the energy released by the combustion of this mass m E_{lib} is consistent with the energy E_{eff} needed to produce the damage.

C9.1.4.1. *Conditions for formation and ignition of ATEX in the building*

C9.1.4.1.1. Propane torch lines involved

We have to assume that the ATEX was formed as a result of a leak due to the failure of an element of the lines that supplied the torches: the hose of each line, as well as its connection to the rigid part by a simple clamp, are certainly the most mechanically fragile elements and the hypothesis of a rupture of the hose or the failure by removal of its fitting is plausible.

It then appears that the maximum quantity of propane likely to have been discharged into the building depends on the type of torch concerned:

– a torch-disassembler was powered under a relative pressure of 2 bar, via the tank's first expansion regulator, which allowed a maximum flow rate of 100 kg/h;

– on the contrary, a welding torch was powered under a relative pressure of 37 mbar, via the wall-mounted device that provided a second trigger and allowed a maximum flow rate of only 1.3 kg/h;

– since the circumstances of the explosion established that the leak lasted no more than 2 h 30 min, the maximum mass of propane likely to have been discharged is 250 or 3 kg, depending on whether the leak resulted from a failure of the hose supplying a torch-disassembler or welding, respectively.

It is therefore only a failure of the hose supplying a torch-disassembler that can be considered.

C9.1.4.1.2. Estimating the mass of propane in ATEX

The mass of propane contained in the ATEX represents only a portion of the total mass of propane discharged: based on a comparison of tank fill rates before and after the explosion, the mass of propane discharged can be estimated to several hundreds of kilograms.

In order to more precisely specify the mass of propane contained in the ATEX, it is necessary to take into account both:

– the geometric conditions of the leak jet;

– the intensity of the damage.

C9.1.4.1.3. Consideration of leak jet conditions

The formation of ATEX resulted from the discharge of propane into air, in the form of a jet and the volume of ATEX formed depends at least on the following parameters:

– the diameter of the orifice, approximately 10 mm;

– the ejection velocity which, at a flow rate of 100 kg/h, is greater than 100 m/s and is therefore characteristic of a turbulent jet;

– the geometric environment of the jet:

- for a jet that would be discharged away from any obstacle, i.e. discharged into free air, the volume of the ATEX would not depend on time,

- but a jet discharged near the walls of the building necessarily impacts obstacles; moreover, the turbulence of the jet promotes the rapid mixing of propane and air. The volume of the ATEX thus increases continuously with the duration of the leak;

– the duration of the leak is therefore also a parameter which the volume of the ATEX formed depends on.

From these elements, the mass of propane contained in the ATEX has been estimated to be several tens of kilograms.

Consideration of damage intensity

For the choice of model enabling us to deduce the mass of propane contained in the ATEX, from the intensity of the damage, we have taken into account the following elements:

– until the explosion, the ATEX remained essentially confined in the building;

– therefore, the flame did not propagate to the open air and its propagation regime could not be that of a rapid deflagration.

The use of the Multi-Energy method is therefore not justified and that of the TNT equivalent model remains relevant (see section 2.5.1.1).

From the damage mapping, we used the abacus TM5-1300 to determine the TNT equivalent of the explosion:

– this mapping shows a certain anisotropy of the effects, i.e. for a given type of damage, the distance at which this type of damage is observed varies according to the direction;

– this means that we could only estimate the TNT equivalent of the explosion by a relatively wide range of values EqTNT between 35 and 150 kg.

From the TNT equivalent, we can estimate the mass of propane contained in the ATEX, taking into account the following additional elements:

– propane combustion heat of 50.4 MJ/kg is about 10 times the decomposition energy of TNT, i.e. 4.69 MJ/kg;

– the average yield of an explosion involving an ATEX, which was formed without any particular homogenization reason, is admitted as approximately 5–10%;

– in this case, the fact that the ATEX formed in the turbulence of a jet was a major reason for the performance to be higher than the above interval.

Based on these elements, we estimate the mass of propane contained in the ATEX to be between 10 and 50 kg, which is a consistent interval with that deduced from the leak conditions.

Estimating ATEX Volume

Knowing that the propane concentration of a stoichiometric propane–air mixture is worth 4.02% volume, the volume of an ATEX consisting of such a mixture and containing 10–50 kg of propane, is between 130 and 630 m^3.

According to the conditions of its formation, the mean concentration of ATEX was certainly lower than the stoichiometric concentration, and its volume therefore fell within a range of values greater than the above values, i.e. between 300 and 1,000 m^3.

Likely nature and location of ignition source

At the time of the explosion, there was no activity in the building and the possible source of ignition is therefore difficult to determine. However, according to the operator, several electrical equipment were energized and all of it was in the eastern part of the building:

– a switchboard was in the office in the southeast corner;

– electrical radiators were located in the rooms along the east wall, one in the office, one in the dressing room and one in the canteen; according to the operator, at least those in the dressing room and canteen were in controlled operation;

– a refrigerator was in the canteen;

– the premises were equipped with an alarm system, with several infra-red detectors (one in the canteen, one in the dressing room and one in the office) and a central located in the office near the electrical panel;

– there were several electric clocks and wireless phones being loaded onto furniture in the office.

It is likely that the ATEX was ignited by an ignition source from a relay connected to one of these different electrical devices and thus located in the eastern part of the building, opposite to the location of the leak.

This assumes that at the time of ignition, the ATEX had sufficient volume to occupy the entire width of the building.

Ultimately, we can admit that the ATEX occupied the entire southern part of the building and therefore that its volume represented a significant fraction of the total building volume, i.e. 2,400 m^3.

This result is consistent:

– on the one hand, with the estimated value of the ATEX volume;

– on the other hand, with the fact that the southern part of the building was completely destroyed (while some roof elements remained in the northern part).

C9.1.4.2. *Causes of the propane leak*

Both technical and organizational causes are responsible for the propane leak.

C9.1.4.2.1. Technical causes

The entire disassembler torch line was under an effective pressure of 2 bar, including the elastomer hose that connected the wall valve to the torch itself. It can be supposed that an element of this hose became fragile and unable to sustain this level of pressure:

– the hose could have been worn over time by friction on the ground, and it could become fragile, close to a puncture;

– the fitting between the hose and the torch or between the hose and the wall valve could have failed, for example, the hose clamp could break.

C9.1.4.2.2. Organizational causes

According to the operator, the personnel had no requirement to systematically close after work all the wall valves feeding the disassembler torches with propane.

In any case, the investigations carried out the day after the accident confirmed this point since, of the five valves found, two were in the open position.

It was because the wall valve in the system where the leak occurred had been left open that this leak could be supplied with propane under 2 bar of pressure, at a flow rate of 100 kg/h, and allow the formation of a large volume of ATEX.

Lessons learned from the investigation

An analysis of the risks of formation and ignition of an ATEX in the facilities would have revealed that a leak on the propane network could have catastrophic consequences, given the high leak rate.

It would have demonstrated that appropriate measures were required to prevent an explosion by:

– strengthening the hoses of the propane supply system of disassembler torches and welding torches;

– adopting instructions to prevent this part from being under pressure during periods when the supervision of facilities is not ensured.

Case 10

Explosion in a Refinery

C10.1. Case presentation

On November 9, 1992, an explosion devastated the refinery processed by TOTAL in La Mède (F – Bouches-du-Rhône).

This accident must be mentioned here because it is a manifestation of the phenomenon of VCE (see section 1.2.3.4), which involves the rapid deflagration of a very large volume of ATEX under semi-confined conditions.

The detailed information obtained from the various investigations carried out in connection with this accident can be consulted in the ARIA report no. 3969 (BARPI 2008).

C10.1.1. *Description of facilities and circumstances of explosion*

The explosion occurred in a catalytic cracking unit:

– a pipe 200 mm in diameter, connected to one of the distillation columns of this unit, contained a liquid effluent at 40°C below 10 bar, consisting of a mixture of propane, butane and propylene;

– a breach of approximately 40 cm in length opened on a horizontal section of this pipe, heavily corroded on its lower generatrix, allowing the discharge of approximately 12 T of effluent in about 10 minutes;

– the effluent discharged was a mixture of liquefied, hot and highly volatile gases that vaporized quickly and mixed with air to form an ATEX cloud

For a color version of all figures, in this case study, see www.iste.co.uk/chaineaux/mechanisms.zip.

of approximately 70,000 m^3, 4–6 m thick, covering an area of approximately 14,000 m^2;

– the cloud drifted and, at a distance of almost 100 m from the breach, its end reached a furnace in the unit where it ignited.

C10.1.2. *Flame propagation regime*

The flame propagated through the cloud, accelerating continuously, so that it was possible to identify three phases during this acceleration:

– the flame first went back to the point of leakage, at a speed that remained moderate (no more than 10 m/s), because there were no obstacles impeding the expansion of the burned gases, and therefore, no significantly strong aerial wave was generated;

– then the flame reached an area where there were obstacles to the flow of fresh (unburned) gases (layers of pipes, control room, distillation columns, platforms, etc.): vortices appeared in these gases and caused a strong acceleration of the flame, which increased to the speed of a rapid deflagration (up to more than 100 m/s), and generated a strong aerial wave (several hundred mbar peak pressure);

– the flame continued to accelerate, reaching at least locally, the characteristics of a detonation (velocity up to 300 m/s and peak pressure up to 2 or 3 bar).

C10.1.3. *Effects of explosion*

The damage caused by the explosion was considerable:

– six fatalities and 37 injuries were reported;

– the very high level of overpressure caused considerable mechanical damage to all types of structure (see Figure C10.1), up to several 100 meters from the center of the explosion (e.g. generalized glass breakage was observed within 1 km);

– the control room, which was severely damaged (see Figure C10.2), was the location of three of the killed people;

– based on the analysis of the observed damage (peak overpressure of the wave as a function of the distance to the center of the explosion), the Eq_{TNT} was estimated to be 5 T, and based on the estimated amount of fuel contained in the ATEX that exploded, 5–10 T, the yield of the explosion could be estimated between 5 and 10% (these figures take account of the fact that the combustion heat of the hydrocarbons involved is close to 10 times the decomposition heat of TNT);

– the explosion ruptured many pipes and ignited the flammable liquids or gases which have been spilled, igniting many violent fires.

Figure C10.1. *View of the devastated facilities*

Figure C10.2. *View of the control room*

Lessons from the investigation

The lessons learned from the explosion are mainly relative to the measures to be taken and the means to be put in place to improve the safety of refining sites against the risks of explosion:

– construction and localization criteria for control rooms;

– strengthening corrosion prevention rules, in particular through regular monitoring of the condition of pipes, at a sufficient frequency;

– improving the effectiveness of certain risk prevention measures such as the number, location, calibration and reliability of gas detectors.

Case 11

Explosions in Recovery Facilities for Cupola Gases

C11.1. Case presentation

C11.1.1. *Description of the facilities*

An explosion occurred in a facility producing mineral wool, which consisted essentially of the two principal elements (see Figure C11.1):

– A cylindrical vertical-axis cupola for continuous melting of vitrifiable materials (blast-furnace slag or basalt): this is obtained by heating to more than 1,000°C, with the heat required being obtained from the combustion of coke in air slightly enriched with oxygen. The materials to be melted and coke are introduced in successive layers at the top of the cupola, while the combustion air, called the "hot wind" is strongly heated upstream, before it is introduced at the bottom of the cupola. The molten materials are continuously drawn from the cupola, through a bottom hole. The combustion of the coke produces CO_2 and H_2O, but this combustion is not complete, so the cupola produces not only these inert gases, but also "cupola gases" (CGs), containing mainly residual flammable gases such as CO and H_2 in small proportions and a large proportion of non-flammable mineral dust otherwise.

– A dust removal of the CGs and a combustion unit for the flammable part of them: they get out through the side wall of the cupola and pass successively through a cyclone, an air cooler, a bag filter and a heat exchanger, where they are heated before they enter a post-combustion chamber, heated by natural gas. This chamber is equipped with a second heat exchanger used for the heating of the hot wind. An exhaust fan located between the filter and the post-combustion chamber maintained the CG flow, which is regulated according to the pressure in the cupola, which is slightly higher than the atmospheric pressure.

The function of the cooler upstream of the filter is to limit the temperature of the filter bags, which must not exceed 135°C. In order to remove the dust deposited on the bags, a fan periodically blows a cleaning air jet through each of them by means of a nozzle moving on a trolley, so that the CGs are slightly diluted by air in the filter.

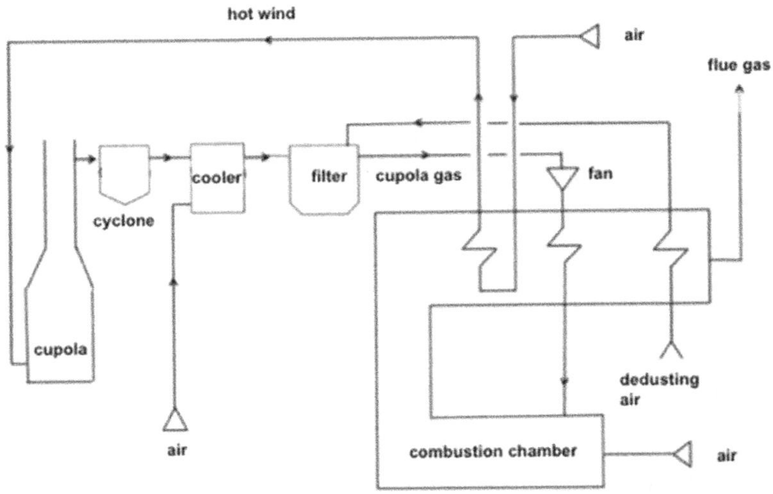

Figure C11.1. *Scheme of the facility*

C11.1.2. *Circumstances of the explosion*

The introduction of the hot wind at the bottom of the cupola was interrupted 5–10 minutes before the explosion. During this time interval, the CGs were admitted into the post-combustion chamber and the CG temperature upstream the post-combustion chamber increased from 300 to 1,000°C within three minutes, kept constant at 1,000°C for two minutes, before lowering more slowly.

C11.1.3. *Explosion damage*

Apart from the deformation of the inspection doors on the cooler casing and the displacement of a soft sleeve located at the inlet of the fan, it is the filter which suffered the main damage:

– deformation of separative metal sheets inside it;

– lifting of circular buffers at the top (dusty gas side);

– combustion of the filter bags and of a sealing material in which the trolley moved.

C11.1.4. *Determination of the explosion mechanism*

The combustible products likely to be involved in the explosion are:

– the CGs flowing from the cupola to the post-combustion chamber;

– a deposit of solid sulfur, present in the filter, both on the dusty side and on the dusted off side of its inner walls, and also on the bags.

C11.1.5. *Flammability of the CGs involved in the explosion*

According to the operator, the CG components are N_2, CO_2, O_2, H_2, CO, H_2S and SO_2. Depending on the current cupola loading operation (coke or vitrifiable products), and depending on the location of the installation considered between the cupola and the post-combustion chamber, taking into account the various air intakes, in particular in the filter and the consequent dilution of the CGs by air, the contents expressed in percentage volume vary between the following intervals:

$13 < CO_2 < 15;$

$6 < O_2 < 12;$

$4 < CO < 8;$

$1.2 < H_2 < 1.8;$

$H_2S + SO_2 < 0.05.$

Moreover, assuming that the H_2S and SO_2 contents are negligible, the N_2 content is expressed in percentage volume:

$63.2 < N_2 < 75.8.$

To characterize their flammability in air, the CGs must first be considered as mixtures of an oxidizer (air), flammable gases (CO and H_2) and inert gases (mainly N_2 and CO_2). However, the above content intervals do not show air as one of the components.

Any O_2–N_2 mixture can be considered as an air–N_2 mixture, the air content of which is deduced from the proportions of O_2 and N_2 in air, which are 20.95% volume and 79.05% volume with a residual N_2 content equal to 100% of its air content.

In Table C11.1, the compositions of the two CG mixtures A and B are given in the first and second columns, respectively:

– mixture A: its O_2 content is maximum, as are its H_2 and CO contents, so its inert content is minimal and equal to 63.2% volume;

– mixture B: its O_2 content is minimal, as are its H_2 and CO contents, so its inert content is maximum and equal to 75.8% volume.

Component of the CGs	Composition of the mixture A (% volume)		Composition of the mixture B (% volume)	
	If air is not considered as a component	If air is considered as a component	If air is not considered as a component	If air is considered as a component
CO	8	8	4	4
H_2	1.8	1.8	1.2	1.2
CO_2	15	15	13	13
N_2	63.2	17.9	75.8	53.1
O_2	12	–	6	–
air	–	57.3	–	28.7

Table C11.1. *Compositions of mixtures A and B of CGs, where air is or is not considered as a component*

The values given for the compositions of mixtures A and B show that:

– their air contents are comprised between 28.7% and 57.3% volume;

– the sum of their CO and H_2 contents is comprised between 5.2% and 9.8% volume;

– the sum of their inert gas contents (CO_2 and residual N_2) is comprised between 32.9% and 66.1% volume.

Therefore, the CGs can be considered as a mixture of air, flammable gases (CO and H_2) and inert gases (essentially N_2 and CO_2), and to characterize their flammability, a triangular representation of this ternary system, described in section 1.1.1.9.1, may be used.

C11.1.5.1. *Flammability of a mixture of one or more flammable gases in the presence of one or more inert gases*

It is known that the operation of chemical processes in which a hydrocarbon fuel is burned in under-oxygenation conditions (such as blast furnaces and cupolas, or incineration furnaces when combustion is not properly regulated) produces different gases, some of which are flammable (H_2 and CO, for the main ones) and others are inert (CO_2 and H_2O for the main ones) and also contain N_2, as well as a few percent of residual O_2.

When an accidental explosion occurs in this type of process, the question may arise whether it may have resulted from the ignition of an ATEX, formed by air dilution of a mixture consisting of one or several of the following seven gases: H_2, CH_4, CO, O_2, N_2, CO_2 and H_2O.

To answer this question reliably, two conditions must be fulfilled:

– the composition of the atmosphere in which the contents of each gas vary, according to the operating conditions of the installation, must be known as precisely as possible;

– the flammability domain of a mixture of flammable gases H_2, CH_4 and CO, in the presence of a mixture of inert gases CO_2, H_2O and N_2, must also be known.

As the first condition can be easily fulfilled if the contents of the different gases are measured by means of analyzers located at different places in the process, the second condition is not easily fulfilled, because no flammability domain has been determined experimentally for a defined ternary system associating all of the following gases:

– a mixture of flammable gases H_2, CO and CH_4, having a defined composition;

– a mixture of inert gases N_2, CO_2 and H_2O, having a defined composition.

Before the question arose in chemical processes, it arose in coal mines, where the above gases are produced when a mass of coal starts a fire in the confined environment of a mine gallery and where it is important to know how to extinguish this fire, taking into account the composition of the mixture of the gases emitted, while avoiding their dilution by air that leads to the explosion of the ATEX formed:

– in its Bulletin 503, the US Bureau of Mines (1952) published a method for characterizing the flammability of a mixture consisting of all or part of the above mentioned gases, as well as the possibility of obtaining a flammable mixture by air dilution of the initial mixture;

– this "associations" method uses the data from Figure C11.2, which shows, in ordinates, the variations of LEL and UEL, for each couple associating a flammable gas and an inert, and in abscissa the ratio of contents (inert/flammable gas) of each pair;

– it should be noted that Figure C11.2 is based on flammability limit values which were determined by means of apparatus and according to protocols different from those of the current standards; these differences explain that the values of the LEL and UEL of flammable gases alone differ slightly from those commonly accepted today, but this difference has no significant effect on the accuracy of the determinations.

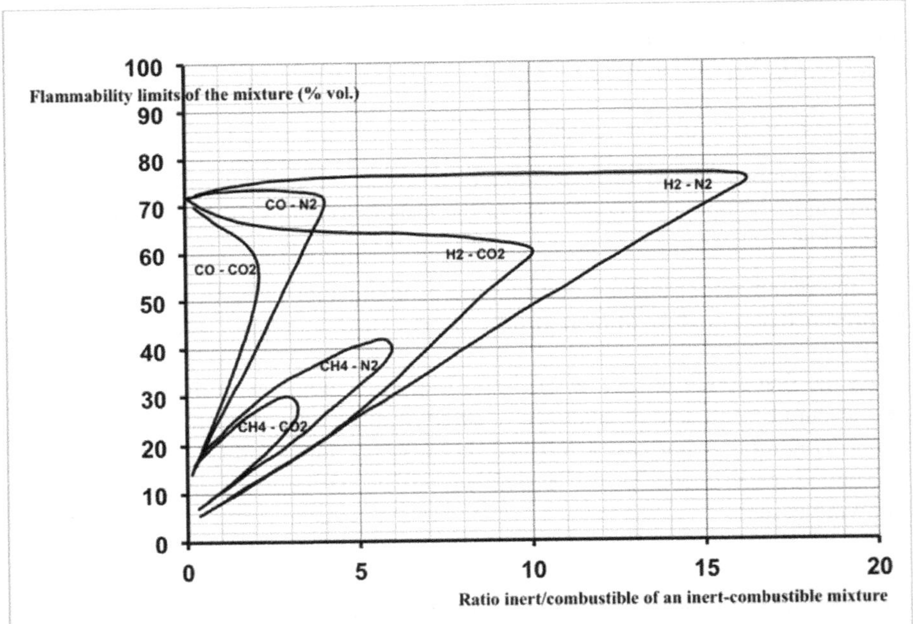

Figure C11.2. *Flammability limits of various combustible gases in the presence of various inert gases*

For example, this method has been applied to a mixture D with the composition given in Table C11.2.

Type of gas	Composition of the mixture D (in % volume)					
	Flammable			Inert		
	H_2	CH_4	CO	CO_2	N_2	H_2O
Content in the mixture D	37.2	30.7	2.9	10	6.7	12.5
LEL	4	4.3	13.7	–	–	–
UEL	75.4	16.3	70.2	–	–	–

Table C11.2. *Composition of the mixture D*

C11.1.5.2. *Case where the mixture contains the inert H_2O*

In order to treat the case of the mixture D that contains the inert H_2O, and although Figure C11.2 does not provide data for H_2O, it is necessary to take into account that H_2O has an inert efficiency comprised between those of N_2 and CO_2. This results in the fact that any flammability domain corresponding to the inert H_2O has an intermediate surface to those of the domains corresponding to N_2 and CO_2. In order to have a conservative approach, an H_2O-free mixture with an N_2 content equal to the sum of the N_2 and H_2O contents of the initial mixture must therefore be considered.

C11.1.5.3. *Associations of flammable and inert gases*

The determination of the flammable properties of the mixture D needs to:

– combine flammable and inert gases;

– determine the explosion limits of each associated gas mixture;

– apply Le Châtelier's law to all associated gas mixtures.

Given in the first part of Table C11.3 are:

– the composition of the fictitious mixture E, which consists of the combination of the whole quantity of H_2 and CO_2 gases contained in the mixture D;

– the composition of the fictitious mixture F, which consists of the combination of the whole quantity of CH_4 and N_2 gases contained in the mixture D. As stated above, the N_2 content was taken as the sum of the N_2 and H_2O contents;

– the CO content of the mixture D, as well as its LEL and UEL (this gas has not been associated with any inert gas);

– the values of the ratio (inert/combustible) corresponding to each fictitious mixture E and F;

– the LEL and UEL values for each fictitious mixture E and F, obtained from Figure C11.2, based on the corresponding ratio (inert/combustible) value;

– the application of Le Châtelier's law to the mixture of fictitious mixtures E and F and to CO, in the proportions of mixture D, leading to the following values of the explosion limits of this mixture:

LEL = 5.7% volume;

UEL = 31% volume.

Given in the second part of Table C11.3 are:

– the composition of the fictitious mixture G, which consists of the combination of the whole amount of H_2 and N_2 (+ H_2O) gases contained in the mixture C;

– the application of Le Châtelier's law to the mixture of fictitious mixtures E and F and to CO, in the proportions of mixture D, leading to the following values of the explosion limits of this mixture:

- the composition of the fictitious mixture H, which consists of the combination of the whole amount of the CH_4 and CO_2 gases contained in the mixture C,

- the values of the ratio r (inert/flammable gas) corresponding to each fictitious mixture,

- the LEL and UEL values for each fictitious mixture G and H obtained from Figure C11.2, are based on the corresponding ratio r (inert/flammable gas) value.

The application of Le Châtelier's law to the mixture of fictitious mixtures G and H and to CO, in the proportions of mixture D, leads to the following values of the explosion limits of this mixture:

LEL = 6.4% volume;

UEL = 34% volume.

To associate an inert with a flammable gas, it is found that the choice of the inert and the flammable gas is indifferent, as well as that of the ratio r of the inert to the flammable gas, since each set of fictitious mixtures leads to relatively similar LEL and UEL values, but this assumes, however, that the fictitious (inert-flammable gas) mixtures all have a flammability domain in air.

Associations of gases	Content in the mixture D (% volume)	Inert/flammable gas ratio r	LEL (in % volume) corresponding to the ratio inert/flammable gas r (given by Figure C11.2)	UEL (in % volume) corresponding to the ratio inert/flammable gas r (given by Figure C11.2)
Fictitious mixture E ($H_2 + CO_2$)	37.2 + 10 = 47.2	CO_2/H_2 = 10/37.2 = 0.27	4	73
Fictitious mixture F ($CH_4 + N_2$ (+H_2O))	30.7 + 6.7 (+12.5) = 49.9	N_2/CH_4 = 19.2/30.7 = 0.62	9	20
CO alone	2.9	0	12.5	72
Fictitious mixture G ($H_2 + N_2$ (+H_2O))	37.2 + 6.7 + (12.5) = 56.4	N_2/H_2 = 19.2/37.2 = 0.52	6	73
Fictitious mixture H ($CH_4 + CO_2$)	30.7 + 10 = 40.7	CO_2/CH_4 = 10/30.7 = 0.32	8	19
CO alone	2.9	0	12.5	72

Table C11.3. *Compositions, LEL and UEL of fictitious mixtures D, E, F, G, H and CO alone*

It appears that, for both LEL and UEL of mixture D, the difference between the values obtained from the fictitious mixtures E and F and those obtained from the fictitious mixtures G and H is small, if the uncertainties inherent in the law of Le Châtelier are considered.

With regard to these associations, it can be seen that:

– the LEL and UEL values for each fictitious mixture G and H obtained from Figure C11.2, are based on the corresponding ratio r (inert/flammable gas) value;

– we can have a single flammable gas, without associated inert, and the LEL and UEL are then those of this pure gas (this is the case of CO in the previous example); on the contrary, the associations must not leave an inert alone, because this would mean that the initial mixture cannot become flammable by air dilution;

– if, as a result of these associations, a fraction of an inert which cannot be associated with a flammable gas remains, it can be said that the initial mixture is not flammable.

C11.1.5.4. *Application to the flammability of the CGs*

Table C11.4 gives the compositions of the CG mixtures (A and B) which are already given in Table C11.1, and which have the following characteristics:

– mixture A: its O_2 content is maximal, as are its H_2 and CO content, so its inert content is minimal;

– mixture B: its O_2 content is minimal, as are its H_2 and CO content, so its inert content is maximal.

<table>
<tr><th rowspan="3">Name of mixture</th><th colspan="5">Type of gas</th></tr>
<tr><th colspan="2">Flammable</th><th colspan="2">Inert</th><th>Oxidizer</th></tr>
<tr><th>H_2</th><th>CO</th><th>CO_2</th><th>N_2</th><th>O_2</th></tr>
<tr><td rowspan="2">Mixture A</td><td rowspan="2">1.8</td><td rowspan="2">8</td><td rowspan="2">13</td><td>65.2</td><td>12</td></tr>
<tr><td>N_2 (when the N_2 part of air has been fictitiously withdrawn):
19.9</td><td>Air: 57.3</td></tr>
<tr><td>Mixture A’
(free of air)</td><td>4.3</td><td>18.7</td><td>30.4</td><td>46.6</td><td>–</td></tr>
<tr><td rowspan="2">Mixture B</td><td rowspan="2">1.2</td><td rowspan="2">4</td><td rowspan="2">15</td><td>73.8</td><td>6</td></tr>
<tr><td>N_2 (when the N_2 part of air has been fictitiously withdrawn):
50.6</td><td>Air: 28.6</td></tr>
<tr><td>Mixture B’
(free of air)</td><td>1.7</td><td>5.7</td><td>21.2</td><td>71.4</td><td>–</td></tr>
</table>

Table C11.4. *Composition of fictitiously A’ and B’ mixtures, If air is not considered as a component*

For each of the mixtures A and B, the method described above has been applied.

Table C11.4 gives the compositions of the CG mixtures (A and B) which are already given in Table C11.1, and which have the following characteristics:

– the air content of mixture A is 57.3%, so we can define an air-free mixture A' which represents 42.7% volume (= 100 – 57.3) of the initial mixture A, and which consists of;

– mixture A: its O_2 content is maximum, as are its H_2 and CO content, so its inert content is minimal:

H_2 = 4.3% volume;

CO = 18.7% volume;

CO_2 = 30.4% volume;

N_2 = 46.6% volume;

– the application of the "associations" method leads to the LEL and UEL values of the mixture in air, respectively:

LEL = 45% volume;

UEL = 69% volume;

– similarly, the air content of mixture B is 28.6%, so we can define an air-free mixture B', which represents 71.4% volume (= 100 – 28.6) of the initial mixture B, with the following composition:

H_2 = 1.7% volume;

CO = 5.7% volume;

CO_2 = 21.2% volume;

N_2 = 71.4% volume;

– unlike mixture A', the calculation shows that mixture B' is non-flammable because it contains too high inert N_2 and CO_2 contents, relative to its flammable gas contents CO and H_2 and therefore has no flammability domain in air;

– for greater clarity, the results were plotted on the ternary diagram of flammable mixtures (CO-H_2), (CO_2-N_2) and air (see Figure C11.3):

- one of the vertices of the triangle represents the air,

- the upper top represents a mixture of CO and H_2 gases, the CO/H_2 proportion of which is assumed to be close to 4, that is, intermediate between the CO/H_2 proportions of mixtures A and B (4.4 and 3.3, respectively),

- the LEL of this CO–H_2 mixture is 72%, since the UEL of H_2 and CO are themselves equal to 72% volume (see Figure C11.2), and the representative point of the CO–H_2 mixture (free of inert) at the UEL was plotted on the diagram,

- by applying Le Châtelier's law, the calculation shows that the LEL of a CO–H_2 mixture, with a CO/H_2 ratio equal to 4, is 8.8% and the representative point of the CO–H_2 air mixture (free of inert) at the LEL has been plotted on the diagram,

- the right summit of the triangle represents a mixture of inert N_2 and CO_2 such that the N_2/CO_2 ratio is close to 5,

- the flammability domain of mixtures (CO–H_2), (CO_2–N_2) air such as CO/H_2 = 4 and N_2/CO_2 = 5 is not known precisely, but it is necessarily close to the flammability domain of mixtures CO–N_2–air that is presented in Figure C11.3 with three differently colored zones;

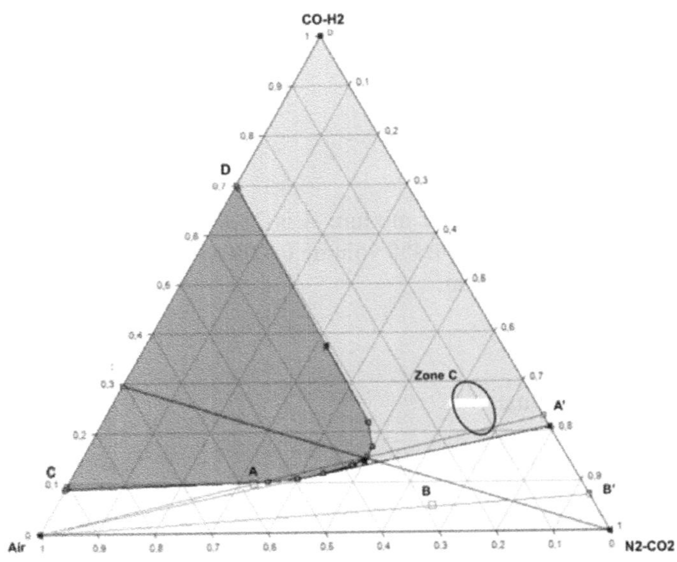

Figure C11.3. *Estimated triangular flammability diagram of the ternary CO–H2–air–N2–CO2 system (such as CO/H2 = 4 and N2/CO2 = 5), at 20°C under 1 bar*

– if the colors of the three zones are considered, it can be said that:

- the dark grey zone corresponds to the mixtures (CO–H_2)–(N_2–CO_2)–air, which are flammable, although they are not diluted by air,

- the light grey zone corresponds to the mixtures that are not flammable, but could become flammable for a certain dilution by air,

- the white zone corresponds to the non-flammable mixtures, which could never become flammable, whatever their dilution by air;

– since N_2 is associated with CO_2 (which is an inert gas more efficient than N_2) and since CO is associated with H_2, whose LEL is lower than that of CO (respectively 4 and 12.5%), the flammability domain of mixtures ($CO–H_2$)–($N_2–CO_2$)–air is actually a little less extensive in the zone of the richest inert mixtures;

– on the representative side of the mixtures ($CO–H_2$)–($N_2–CO_2$), the representative points of the mixtures A' and B' were plotted, and on the lines joining points A' and B' at the representative summit of the air, points A and B, the air contents of which are known;

– the air-B' line does not cut the flammability domain (B' is in the white zone), and thus it corresponds to the result of the previous calculation that the mixture B' is not flammable, as well as no mixture resulting from its dilution to air;

– on the contrary, the air-A' straight line cuts the flammability domain (A' is in the light grey zone and point A is just at the limit of the flammability domain).

C11.1.5.5. *Application to the flammability of the CGs*

In the operator's experience, the sulfur dust deposited on the bags and on the inner wall of the filter consists of very low size grains. This corresponds to the fact that this dust results from the condensation of sulfur vapor present in the CGs (see below), this condensation occurring in the filter because its temperature, necessarily moderate to prevent thermal degradation of the bags, is less than the dew point of the sulfur vapor.

In the final analysis, since sulfur dust can be easily blown off, in particular by the cleaning air jet, it is mixed with the gaseous ATEX supposed to be present in the filter, so that the explosion occurs in a hybrid ATEX consisting of a mixture of CGs and sulfur dust.

C11.1.5.6. *Synthesis of the explosion mechanism*

Therefore, it can be concluded that the explosion most likely resulted from the following mechanism:

– after the flow of hot wind has been stopped, the composition of the mixture processed in the filter and up to the post-combustion chamber gradually evolved to an ATEX;

– according to the diagram in Figure C11.3, it was necessary that, starting from the composition of the mixture represented by point A, the CO and H_2 content increased slightly, while the air content decreased slightly;

– this situation occurs if the air flow decreases slightly to the benefit of the CG flow;

– the post-combustion chamber temperature was sufficient to ignite this ATEX;

– at the beginning of the evolution of its composition, the ATEX was not such that, given the velocity of the gas flow at the inlet into the post-combustion chamber, the flame propagated upstream;

– this flame was then maintained for a few minutes in the vicinity of a probe which, at that time, indicated an abnormally high temperature:

- then, the composition of the ATEX continued to evolve gradually and enter further into the flammability domain, allowing the flame to propagate in the pipeline, up to the filter and counter-current of the gas flow,

- this flame propagated through the flow control throttle valve; actually, although it was closed to stop the hot wind flow, a little orifice (approximately 5 mm width) remained, which was sufficient to allow the propagation of a flame through it.

It may seem surprising that the flame propagated upstream of the gas flow whose velocity was greater than its fundamental flame velocity of this flow (a few tens of cm/s); actually, when combustion starts, it generates an overpressure in the gas that is sufficient to stop its flow, and it is ultimately through a quiet gas that the flame has propagated upstream:

– the flame propagated to the fan, producing a sufficient overpressure to move the soft sleeve connecting the line to the inlet of the fan;

– finally, the flame propagated up to the filter, on the dedusted gas side, and it produced an overpressure that we estimate at a few hundred millibars, sufficient to blow out the filter doors; note that it is in a gaseous ATEX that the flame propagated from the chamber, but it is very likely that, at least on the dedusted gas side, and perhaps also in the downstream pipe, the sulfur dust present on the inner walls of these elements was blown off by the explosion resulting from the propagation of the flame, and participated in the explosion itself;

– the explosion in the filter, on the dedusted gas side, also propagated to the sulfur dust present in the dusty gas part and transmitted to that part, in the area closest to the filter inlet, through the sleeves moved by the blast of the explosion or damaged previously by too hot gases;

– an explosion of sulfur dust then developed in the filter on the dusty gas side and produced sufficient overpressure to damage the cooler box closest to the filter;

– the flame probably also began to develop outside of the filter, either through the circular buffers present at its top, or through the orifice in which the trolley moves, but the flame could not propagate upstream of the filter, because the CGs were under-oxygenated there;

– finally, the explosion was followed by a widespread bag fire, which was fed by the sulfur deposited on the bags.

C11.1.5.7. *Other CG recovery facility explosion*

A second case of an explosion, having occurred in a facility used for the production of engine blocks from cast iron, had to be investigated. As in the first case, the CGs passed through a release chamber where they were roughly dedusted in a cyclone and then burned in a post-combustion chamber. This chamber played the role of a heat exchanger, feeding the cupola with combustion air as "hot wind", and produced combustion gases which were dedusted downstream.

For established conditions, the CGs had a composition slightly different from the first case, since they contained only 2–3% volume oxygen (which corresponds to an air content of between 9.5% and 14.3% volume), but more combustible gases (18–25% volume CO and 0–4% volume H_2) and a CO_2 content between 8% and 12% volume.

In Figure C11.3, the zone C corresponds to the variation area of the CG composition. This zone C is inside the light grey zone, which means that its dilution by a sufficient quantity of air causes the representative point of the resulting mixture to move towards the representative pure air point, which causes it to enter the flammability domain, and leads to the formation of an ATEX.

The circumstances of the two explosions are not identical since, in the second case, it was sufficient for air to be sucked into the CGs, forming an ATEX which was ignited in the post-combustion chamber, the flame propagating upstream.

Lessons learned from investigation of explosion in cupola facilities

When the hot wind is stopped, there is a significant risk of ATEX forming in the filter and downstream piping:

– The means of preventing an explosion by continuously analyzing the composition of the gases (oxygen, CO and H_2) admitted into the chamber is in principle possible, but this is a measure that must be combined with an enslaved system that controls the opening of the CG flow control valves and dilution air, and the variability of these flows makes it a poorly reliable means.

– The temperature in the combustion chamber is high enough to constitute a certain source of ignition of an ATEX which would be admitted there: assuming the presence of an ATEX in the part of the installation located upstream of the combustion chamber, it would be possible to prevent the propagation of the flame upstream by means of a flame arrester installed just upstream of the chamber, but this would require a sufficient dust removal of the CGs by means of a filter located between the cupola and the chamber.

– Such a means of prevention must be accompanied by the installation of explosion vents on the upstream capacities (filter, expansion chamber and cyclone), in order to protect them effectively against the effects of an explosion.

– The presence of sulfur in the filter is due to the condensation of sulfur vapor present in the CGs. If it appears difficult to avoid the presence of sulfur in these gases, it may be possible to avoid condensation of the sulfur vapor they contain; this assumes, however, that all components of the CG filtration system are at a temperature above the dew point of the sulfur vapor, and in particular the filter bags. In the first investigated case, the maximum allowable temperature of the bags was 135°C, which was very close to the dew point of the sulfur vapor. Then, the conditions for condensation of sulfur vapor on the filter bags were fulfilled.

– During the explosion, the flame propagated upstream of the pipe, despite the fact that the throttle valve was not fully closed, and that the CGs were discharged through a remaining little orifice (approximately 5 mm width): it is not surprising that the hot gases generated by the flame create an overpressure in the pipe; this overpressure is sufficient to stop the flow and the flame can then propagate through a gas whose flow has been stopped. It is therefore not possible to estimate a sufficient speed of an ATEX flowing in a pipe to prevent the propagation of a flame upstream.

Case 12

Explosion of Acetone Vapor

C12.1. Case presentation

A very violent explosion, which occurred in an electrical substation, had to be investigated.

C12.1.1. *Description of the facilities*

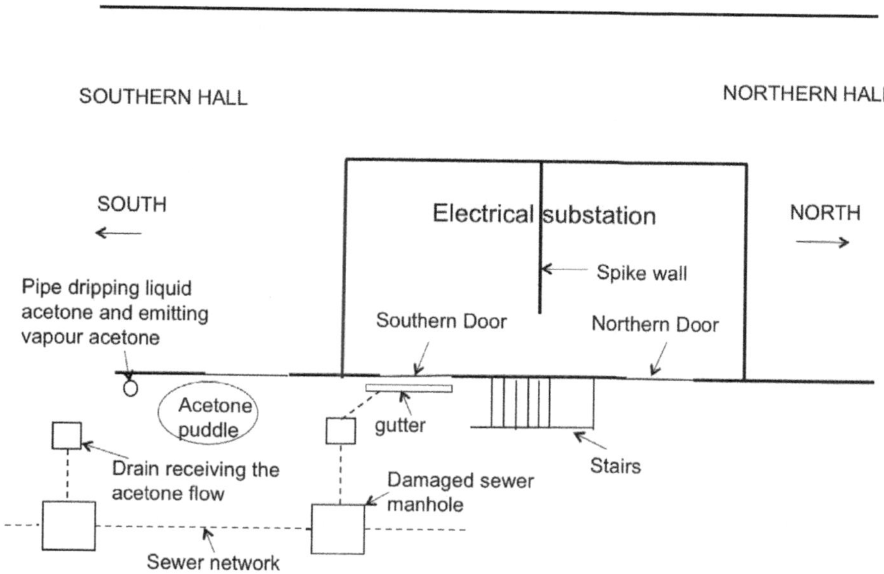

Figure C12.1. *Scheme of the explosion site*

The electrical substation, rectangular in shape (volume approximately 60 m^3), was located on the ground floor of a building comprising two contiguous halls where chemical reactors were operated (see Figure C12.1). This substation was arranged on a North–South axis, and its outer wall constituted the large side of the rectangle and comprised two metal doors (2 m^2 for each), a few meters apart and separated by a staircase leaning against the outer wall. Only the southern door was equipped, in its lower part, with a ventilation grid (surface 0.3 m^2). A spike wall, discontinued and perpendicular to the western one, separated the substation into two roughly equal volumes. The ceiling slab of the substation was made of concrete.

There was no communication between the halls and the substation, which only communicated with the outside through its two doors.

C12.1.2. *Circumstances of the explosion*

While some syntheses were in progress in both halls, under acetone reflux, the explosion occurred approximately five minutes after an operator, who was outside in the vicinity of the substation, made the following findings:

– He smelled a strong odor of acetone in the vicinity of the doors, which were closed.

– He observed an acetone puddle (about 0.7 m^2 area) near the south door. This puddle was fed by a little stream of acetone, dripping from the end of a vertical pipe, which came out at the foot of the outer wall of the south hall and was connected to the condenser vent of a reflux reactor.

– Finally, he found that liquid acetone was flowing from the puddle to a nearby sewer manhole, which was located in the immediate vicinity of the south door and connected to several others by the sewer network.

C12.1.3. *Description of explosion damage*

The explosion severely damaged the substation:

– each of the two doors and their frames were torn off and thrown several meters away;

– similarly, around the doors, parts of the outer wall, made of cellular concrete bricks, were blown out and debris was thrown several tens of meters away;

– the other three solid block walls were broken up, fragmented and pushed back to the halls and, in their movement, deformed the facilities that were attached to them (pipes, electrical wires, etc.);

– one part of the spike wall was blown and its debris was projected northward while remaining inside the substation;

– finally, the reinforced concrete ceiling slab cracked and partially collapsed.

Some elements of the sewer network were also damaged:

– the masonry joint, which connected the sewer pipe to the closest manhole to the south door, was broken;

– moreover, the cast iron cover that closed this manhole, as well as those closing more distant manholes, were projected far away from there.

Some electrical cabinets had their doors thrown away and sheets of paper, which were fixed to the inner side of these doors, had their edges marked by burned black traces. No other significant thermal effect, either in the substation or outside, has been observed.

C12.1.4. *Mechanism of the explosion*

The mechanism of the explosion has been determined from its circumstances and the effects it produced:

– the acetone coming out of the end of the pipe was partly liquid, dripping on the ground, and partly vapor, mixing with air, either in the immediate vicinity of the pipe or above the puddle (see Figure C12.1);

– an ATEX was formed locally, which was pushed by the wind, at a distance of approximately 2 m, up to the ventilation grid of the south door of the substation, through which it was sucked into it;

– the ATEX was ignited by an electrical spark produced during the normal operation of one of the switchboard components (electromechanical relay, contactor, etc.), and the flame developed throughout the substation;

– the flame also spread to the ATEX present outside the substation, up to the closest manhole to the south door, and ignited the ATEX that was formed inside, which developed a sufficient overpressure to project the manhole cover and break the joint with which the sewer pipe was connected to the manhole.

For the covers of the most distant manholes, there are a priori two possible scenarios:

– on the one hand, the explosion propagated to the ATEX that was present within the sewer network and the overpressure generated inside it was sufficient to lift and throw away the covers; the covers were sucked by the depression phase of the aerial overpressure wave produced by the explosion; it is certainly the first scenario which is the most probable for the closest cover to the south door, since in addition to the projection of the cover, the internal masonry joint of the manhole was broken;

– on the other hand, in the absence of ruptures observed in the other manholes, the second scenario is also plausible for their covers, especially since it is quite common in this type of configuration: the depression phase following the overpressure phase of the aerial wave usually reaches an amplitude of approximately 10 mbar, which is sufficient to lift and project, for example, a cover of 0.5 m x 0.5 m, weighing 25 kg.

The different masses of acetone, in liquid or vapor state, involved in the mechanism of the explosion, have been considered:

– m is the total mass of acetone, in liquid or vapor state, which was discharged by the pipe up until the explosion occurred (from an estimation by the operator, m was comprised between 50 and 100 kg);

– m_l is the mass of liquid acetone flowing to the puddle and then to the drain of the sewer system;

– m_2 is the mass of acetone in vapor state discharged by the pipe;

– m_3 is the mass of acetone in vapor state which has been evaporated from the puddle;

– m_4 is the mass of acetone in vapor state which could be present outside the substation and which has been partially sucked into it;

– m_5 is the mass of acetone in vapor state which was sucked into the substation;

– m_6 is the mass of acetone in vapor state which was present in the ATEX which exploded.

The masses m, m_1, m_2, m_3, m_4, m_5 and m_6 are linked by the following relationships:

$$m = m_l + m_2;$$

$$m_4 = m_2 + m_3;$$

$$m_5 < m_4;$$

$$m_6 < m_5.$$

To confirm the above mechanism, it is necessary to verify that m_6 and m have consistent values.

C12.1.4.1. *Estimation of m6 from the intensity of damage*

An estimation of m_6 has been based on the maximum overpressure reached in the substation, according to the nature and intensity of the damage:

– The two doors of the substation were the least resistant elements of its structure. They were torn up and projected at about 10 meters when the overpressure developed there reached a value of between 40 and 80 mbar.

– After the doors broke, the overpressure continued to increase to a maximum value P_{exsub}, which can be estimated at 200 mbar; this value was sufficient to blow the east wall of the substation, throwing its debris up to several tens of meters, and produce the damage observed on the ceiling slab and the other walls. The nature and the level of destruction, as well as the value of P_{exsub}, correspond to a very violent explosion.

– A comparison between P_{exsub} and the maximum overpressure P_{ex} = 8 bar, which would be developed by the explosion of a stoichiometric ATEX contained in a resistant closed container (P_{ex} is similar to that shown in Table 1.4, for a stoichiometric air–butane ATEX), corresponds to the ratio $P_{ex}/P_{exsub} = 40$.

– If a volume v of a stoichiometric acetone ATEX is supposed to explode in the whole volume V = 60 m^3 of the substation, it is for v = V/40 = 1.5 m^3 that the explosion would develop approximately the same overpressure P_{exsub} = 200 mbar.

– Since the concentration of acetone in a stoichiometric ATEX is equal to 5% volume, the mass m_6 of acetone contained in the volume v was close to 200 g.

When attempting to determine the mass m_5 of acetone in vapor state present in the substation, the following have to be considered:

– it has already been pointed out in section 2.7.2 that, when an ATEX explodes in a building with a conventional and little resistant structure, it is not possible to accurately estimate the mass of the fuel present in the ATEX;

– moreover, it must also be taken into account that the air–acetone mixture was not homogeneous throughout the substation and that it is only a little part of this mixture which was an ATEX, i.e. had an acetone content comprised between its explosion limits 2.6% and 9.3% volume;

– ultimately, it is certain that m_5 was higher than m_6 and possibly reached 0.5–1 kilo.

C12.1.4.2. *Estimation of m6 from the values of m, m_1, m_2, m_3 and m_4*

The ATEX formed in the substation contained acetone vapor given by the following two sources:

– the vapor evaporated from the puddle;

– the vapor directly discharged by the pipe.

C12.1.4.3. *Calculation of the mass m_3 of acetone evaporated from the puddle*

The value of m_3 has been calculated with an evaporation model presented in section 1.1.1.5.3, which is based on the following relationship [C12.1]:

$$q(Ts) = 22 . S . V . M^{0.4}. P_{sat}(Ts)/P_{atm} \qquad [C12.1]$$

where:

– q(Ts) is the evaporation rate of acetone from the puddle (in g/s);

– M is the molar mass of acetone, i.e. M = 0.058 kg/mole;

– S is the puddle area (in m^2);

– V is the air velocity at the puddle surface (in m/s);

– T_{eb} is the boiling temperature of acetone under atmospheric pressure, i.e. 56°C;

– T_s is the temperature of the acetone on the surface of the puddle (in °C);

– $P_{sat}(T_s)$ is the saturated vapor pressure of acetone at T_s (in Pa);

– P_{atm} is the atmospheric pressure, i.e. 101,300 Pa.

The S, V and $P_{sat}(T_s)$ values were selected, on the basis of the following:

– the operator indicated the maximum area S of the puddle, i.e. 0.7 m^2, but they were unable to specify the time limit after which this value was reached;

– the outer wall of the substation was sheltered from a weak westerly wind and the upper limit of the range of applicability of the model, i.e. V = 1 m/s, has been considered;

– for the temperature T_s of acetone on the surface of the puddle, it must be taken into account that it was fed by a flow of hot acetone, and also that acetone was cooling, both because of its contact with a colder soil and because evaporation is an endothermic phenomenon. Hence, it has been considered that T_s remained equal to 20°C, which corresponds to a value of $P_{sat}(T_s)$ equal to 24,500 Pa.

For the above values of parameters S, V and $P_{sat}(T_s)$, the model leads to the result that the evaporation rate of acetone from the puddle was q = 4 kg/h (or 1.1 g/s).

For such a rate, the mass of acetone evaporated in 1 s by an area of 1 m^2, i.e. 1 g, represents 0.4 dm^3 of vapor. This vapor mixes with air and the volume v of mixture, assumed to be an homogeneous and stoichiometric ATEX, is v = 7.7 dm^3, which corresponds to a very flat square shape, having a base and a thickness equal to 1 m^2 and 8 mm, respectively. Actually, the existence of a concentration gradient on the surface of the liquid suggests that the ATEX was even thinner.

This very low value of v means that the acetone vapor evaporated from the puddle was present only in the immediate vicinity of its surface and did not contribute at all to the formation of the ATEX present in the substation.

C12.1.4.4. *Origin of the ATEX present in the substation*

Since the mass m_3 of acetone evaporated from the puddle is very low, the only possible origin of the ATEX is the acetone vapor, having a mass m_2, which has been discharged through the pipe.

C12.1.4.5. *Comparison of m_2 and m_3*

A comparison between m_2 and m_3 is possible if the following are considered:

– the pipe, from which a biphasic mixture of acetone, in liquid and vapor phase, has been discharged, was connected to the vent of a condenser;

– when a condenser is normally cooled, it is only a weak flow of vapor which is discharged through the vent; but, if the condenser is not sufficiently cooled, a biphasic mixture of liquid and vapor begins to flow out of the pipe at an increasing rate. Simultaneously, the temperature of the pipe wall increases, and the proportion of vapor in the mixture increases too;

– it can be supposed that, within the minutes preceding the explosion, the flow of acetone vapor was important and much higher than the evaporation rate of the puddle.

Therefore, in any case, m_3 was much lower than m, which was certainly not lower than 50 kg. The mass m_4 of acetone in vapor state, which could be present outside the substation and possibly sucked into it, can be considered as similar to the mass m_2 of acetone in vapor state discharged by the pipe.

Ultimately, the mass m_2 of vapor discharged by the pipe was much greater than the mass m_5 of acetone present in the substation, which has been estimated at only 0.5–1 kg.

Then, it results from these estimations that there is a good consistency between the mass of ATEX, which exploded in the substation and the mass of acetone in vapor state, which has been discharged by the pipe and sucked into the substation.

Lessons from investigation of an explosion of acetone vapor

The evaporation rate of a volatile liquid such as acetone, which is spread on the ground in the form of a puddle with a wind of 1 m/s, is about 1 $g/m^2.s$, when the temperature of the liquid is around 20°C:

– under these conditions, the acetone vapor mixes with air in the immediate vicinity of the liquid surface and an ATEX is formed with a very small thickness (less than 1 cm), which does not extend beyond the puddle;

– on the contrary, a stream of acetone vapor, coming out of the orifice of a pipe at a temperature of about 50°C, at low pressure and thus at low speed, mixes with air to form an ATEX whose volume is variable according to the vapor flow;

– this ATEX is gradually diluted in the air, downwind of the orifice, but if this wind is moderate, it can be transported for several meters, without its dilution by the air significantly decreasing its volume.

When an ATEX is present in a building consisting of a structure with low resistance to overpressure, and when the building is severely damaged or even destroyed by the explosion of this ATEX, it is not possible to determine its volume accurately nor, therefore, the mass of fuel it contained.

The explosion could have been prevented if:

– the condenser failure was anticipated (its vent pipe should not have opened at the foot of a wall, but in open space);

– the electrical substation had been kept under a slight overpressure, to prevent an external ATEX from entering it.

Case 13

Explosion of Vapor of Toluene

C13.1. Case presentation

C13.1.1. *Description of the facility and of the circumstances of the explosion*

The mechanism of an explosion having occurred in an IBC (Intermediate Bulk Container) had to be determined. This container was in polyethylene, had a capacity of approximately 1 m^3 and contained about 100 liters of water, with traces of toluene.

At the time of the explosion, the IBC was on a pallet jack and an operator was moving it inside a room. It was equipped with a screwed lid, which had been kept closed for a long time at room temperature.

C13.1.2. *Effects of the explosion*

The damage caused by the explosion was relatively low:

– the IBC was destroyed, as a result of an internal overpressure exceeding its breaking pressure: its upper part was torn along a significant fraction of its perimeter and partially separated from the lower part;

– some elements of its metal frame were twisted;

– the operator was not injured or burned by the explosion.

C13.1.3. *Determination of the mechanism of the explosion*

C13.1.3.1. *Physicochemical characteristics of toluene*

Toluene has the following physicochemical characteristics:

– solubility in water: 515 mg/L;

– molar mass: 92 g/mole;

– water density: 0.9;

– flash point P_E: 4°C;

– explosion limits: LEL = 1.2% volume, UEL = 7.1% volume;

– saturated vapor pressure at 20°C: 2,900 Pa.

C13.1.3.2. *Presence of an ATEX in the IBC*

The explosion resulted from the ignition of an ATEX formed in the IBC, and one of the issues of the investigation was to explain this formation.

The IBC contained liquid and gaseous phases whose characteristics were as follows:

– a liquid-vapor equilibrium had enough time to be established in the IBC at room temperature;

– the gaseous phase, i.e. the "IBC sky", was a mixture of air and toluene vapor whose toluene content corresponded to the saturated vapor pressure at 20°C, i.e. 2,900 Pa. Since this toluene content, i.e. 2.9% volume belongs to the flammability range, this confirms that the IBC sky was an ATEX;

– the liquid phase comprised two parts:

- the main part was a solution of toluene in water,

- a certain quantity of liquid toluene, supernatant at the surface of water. Its presence is a condition for the existence of the equilibrium, and it requires the solution of toluene in the water to be saturated.

According to these assumptions, the distribution of toluene in the IBC was as follows:

– in the volume of the IBC sky, a toluene vapor close to 2.9% volume corresponds to a toluene mass close to 100 g;

– in 100 L of water, the saturation corresponds to a mass of about 50 g of dissolved toluene;

– the mass of the supernatant toluene cannot be precisely known, but a very low one may be sufficient.

It can be concluded that:

– the total toluene mass contained in the IBC was at least 150 g;

– this figure corresponds to a mass concentration of toluene close to 0.15% in the aqueous effluent before it was introduced into the IBC.

According to the operator, such a concentration was consistent with the origin of the aqueous effluent contained in the IBC.

C13.1.3.3. *Influence of water vapor on the ATEX composition*

This case study gives the opportunity to examine the influence of water vapor on the ATEX composition:

– at room temperature, the saturated vapor pressure of water is close to 20 mbar, so the concentration of water vapor in the IBC sky is close to 2% volume;

– in combustion phenomena, water vapor plays the role of an inert gas, but such a low concentration does not affect the explosion characteristics of the IBC sky;

– these characteristics only depend on the concentration of toluene vapor in the IBC sky, and they do not depend on the presence of water in the liquid phase;

– it would be different if the temperature was significantly higher than ambient, because the concentration of water vapor in the sky would increase and the IBC sky would possibly become non-flammable.

More generally, where a liquid-vapor equilibrium can be established in a IBC containing water, at the surface of which there is a flammable supernatant liquid, non-miscible with water and whose flashpoint is less than the ambient temperature, the atmosphere present in the IBC sky is an ATEX, as it would also be if this flammable liquid was pure in absence of water.

So that this equilibrium can be established, the water must be saturated with the flammable liquid. If this condition is satisfied, it is not necessary that a large amount of supernatant liquid is present, but if the quantity of this liquid is sufficient to be observed, its presence indicates the explosive nature of the IBC sky.

C13.1.3.4. *Nature of the ignition source of the ATEX*

Among the sources of ignition that may have triggered the explosion, the only one that seems possible is electrostatic. It could be related to the occurrence of a spark that sprung in the IBC sky, between two components that were at different electrostatic voltages and carried electrostatic charges that were generated as a result of the IBC's displacement in the room or during a previous operation.

However, the phenomenon that led to an increase of the electrostatic voltage of an IBC component isolated from the earth or isolated from another component, and then the phenomenon that allowed the release of the accumulated energy, could not be identified.

Lessons learned from the investigation of an explosion of toluene vapor

ATEX is present in the sky of a container that contains, at room temperature, a flammable liquid whose flash point is less than the ambient temperature, and where a liquid-vapor equilibrium has been established.

It is also an ATEX that is present in the sky of a container containing water at room temperature, at the surface of which there is a flammable supernatant liquid, non-miscible with water, whose flash point is less than the ambient temperature, if a liquid-vapor equilibrium can be established; the amount of supernatant liquid may be very low and may even be zero if the water is saturated, and if the vapor concentration in the sky is convenient.

Case 14

Explosion of Vapor of Kerosene

C14.1. Case presentation

The mechanism of an explosion that damaged a continuous cement production facility and its upstream and downstream parts had to be determined.

C14.1.1. *Description of the facility*

The facility consisted essentially of:

– a cylindrical rotating kiln, 70 m long and 4.6 m diameter, having a slightly inclined axis, which was continuously loaded with limestone in its upstream end;

– a fixed parallelepipedal heating hood, installed upstream of the kiln, and comprising a main burner located on its axis, which was fed by an atomized liquid fuel and a flow of combustion air ensuring this atomization, and which was equipped with a propane pilot flame;

– the downstream part included:

- a unit for solidification of the molten cement flux (clinker) produced inside the kiln,

- a unit for de-dusting the flue gases of the burner, by means of an electro-filter.

C14.1.2. *Circumstances of the explosion*

The explosion occurred during the start-up of the kiln under the following circumstances:

– the kiln was restarted after a maintenance operation that involved a series of burner ignition and extinguishing sequences, to gradually dry the kiln, the refractory coating of which had just been replaced;

– the main burner was then started for a production phase, with kerosene as fuel, and not diesel fuel as usual;

– while the start-up phase was proceeding normally, the main burner shut down due to a maladjustment of the combustion parameters, which the operator attributed to either an excessive kerosene flow or an insufficient atomization;

– whatever the true reason for this extinction, there has been no simultaneous automatic shutdown of the kerosene supply, which is a normal preventive safety action, and this event suggests that a significant amount of liquid kerosene has spread on the sole of the kiln.

The explosion occurred shortly after the burner was extinguished.

C14.1.3. *Description of explosion damage*

The explosion produced only limited material damage, which was mainly relative to the upstream and downstream parts of the kiln:

– upstream of the kiln, the two access doors to the hood were torn off, and other elements were deformed or moved;

– downstream of the kiln, the hinges of the double access doors to the cooling unit were also torn off;

– the welds of the sheet metal ducts that carried the flue gases from the cooling unit to the electro-filter were torn;

– finally, the aerial pressure wave, emitted during the opening of the doors upstream and downstream of the kiln, blew several cladding plates from the kiln building.

C14.1.4. *Determination of the explosion mechanism*

According to the nature of the products used, the explosion necessarily resulted from the ignition of an ATEX, which formed in the upstream part of the kiln.

C14.1.4.1. *Identification of the ATEX components*

A priori, the fuel constituting the ATEX could either be the propane supplying the pilot flame or the kerosene:

– the propane hypothesis has been rejected, since the pilot flame remained on for the duration of the start-up phase until the time of the explosion;

– therefore, the ATEX that exploded was necessarily a mixture of kerosene vapor and air.

C14.1.4.1.1. Physicochemical characteristics of diesel fuel and similar liquids

Based on its flash point specifications, the diesel fuel corresponds to a petroleum distillation cut that is significantly less volatile than kerosene. Moreover, their respective flash points are similar to those of pure products of the series of n-alkanes, i.e. n-nonane, n-decane and n-undecane.

Name of the combustible	Flash point P_E (°C)	Boiling point at ambient pressure (°C)	Auto-ignition temperature (°C)	Molar mass M (g/mole)	Volumic mass (kg/dm³)	LEL (% volume)
Kerosene	$37<P_E<65$	142–282	210	132.4	0.75	0.5 to 0.7
Diesel fuel	> 55	150–380	250–300	NA	0.83–0.88	0.5
n-nonane	31	151	205	128	0.724	0.9
n-decane	46	174	205	142	0.736	0.7
n-undecane	65	196	NA	156	0.739	0.6

Table C14.1. *Different physicochemical characteristics of five liquid fuels*

C14.1.4.1.2. Kerosene evaporation rate

The operator considered that:

– during the successive phases of ignition and extinction of the burner, which preceded the start-up of the kiln, the sole reached an estimated temperature of at least 140°C;

– the main burner extinguished because it discharged an insufficiently atomized jet of liquid kerosene, which spread on the sole in the form of a puddle (such a reason is more probable than a maladjustment of the fuel/air ratio);

– the sole temperature was sufficient for the kerosene to evaporate and mix with the air flowing in the kiln.

Under the accident conditions, a significant volume of ATEX could be formed if the ratio of the evaporation rate of kerosene to the airflow rate was within the flammability range.

C14.1.4.1.3. Modeling of the formation of ATEX by evaporation of a flammable liquid

When the surface of a puddle of flammable liquid is subjected to a horizontal air flow, the vapors emitted by the liquid are drawn and mixed with air. The literature provides several empirical models to determine the conditions required for the air-vapor mixture to form a significant volume of ATEX.

The model published in 2009 by INRS (Chaineaux 2009) has been used to determine the evaporation rate of the kerosene puddle, which depends on different parameters according to the following relationship [C14.1]. This relationship only differs from the [1.3] one, which is justified by the coefficient 5 instead of 22 and by the change in unit of the quantity q, expressed in kg/h instead of g/s (see section 1.1.1.5.3):

$$q = 5\ .\ (P_{sat}(TL)/P_{atm}).\ S\ .\ V.\ M^{0.4} \qquad [C14.1]$$

where:

– q is the evaporation rate of the liquid (in kg/h);

– M is the molar mass of the liquid (in g/mole);

– S is the surface of the liquid puddle (in m^2);

– V is the velocity of air at the surface of the puddle (in m/s);

– TL is the liquid temperature (in K);

– T_{eb} is the boiling temperature of the liquid under atmospheric pressure (in K);

– $P_{sat}(TL)$ is the saturating vapor pressure of the liquid at TL (in Pa);

– P_{atm} is the atmospheric pressure (in Pa).

This relationship is applicable for the following ranges of values:

– TL << Teb (at Teb, evaporation is no longer governed by the vapors being drawn into the air flow and also by their emission under the effect of boiling);

– S up to several m^2 (the flow rate q is proportional to S and, beyond certain dimensions of the puddle, the air is saturated with vapors and the proportionality is no longer respected);

– $0.3 < V < 0.9$ m/s (tests were carried out for V = 0.3, then 0.9 m/s so that interpolation between these values is possible, but extrapolation is more uncertain).

From the value of q, the following relationship makes it possible to determine for which minimum value of the air flow Q, the air-vapor mixture formed downstream of the puddle is sufficiently diluted by air, so it cannot form a significant volume of ATEX.

$$Q = 100 . q . (R\ T/P_{atm}) / (M . LEL) \qquad [C14.2]$$

where:

– Q is the minimum air flow (in Nm^3/h);

– q is the evaporation rate of the liquid (in kg/h);

– R is the gas constant = 8.31 J. K^{-1}. $mole^{-1}$;

– T is the air temperature (in K);

– P_{atm} is the atmospheric pressure (in Pa);

– M is the molar mass of the liquid (in g/mole);

– LEL is the lower explosion limit of the liquid (in percentage volume).

C14.1.4.1.4. Application of the model to the present case

For kerosene, the variation of $P_{sat}(TL)$ as a function of temperature is not known, and n-nonane and n-decane, which have boiling points close to that of kerosene, have been considered.

For the air velocity at the start-up of the kiln of 17 m^2 cross-section, the hypothesis of a maximum air flow of 4,000 Nm^3/h, and a homogeneous velocity field in the right section of the heating hood, leads to a maximum value of V that is a bit lower than 0.1 m/s. However, the velocity field may not be homogeneous and the air velocity could have been greater than 0.1 m/s close to the sole. Then, it can be supposed that it was not very far from 0.3 m/s and that the validity conditions of the model were met.

For the surface of the puddle, it is difficult to give a precise value. The quantity of kerosene, a little over 100 L, which spread on the sole from a liquid jet discharged by the burner positioned at 2.3 m above, during the 10 minutes between its last extinction and the instant of explosion, must be taken into account. Then, S can be estimated to a value between 10 and 20 m^2.

According to the operator's experience, the burner operated for a total of 94 minutes, and this duration was sufficient for the sole to reach or exceed approximately 140°C.

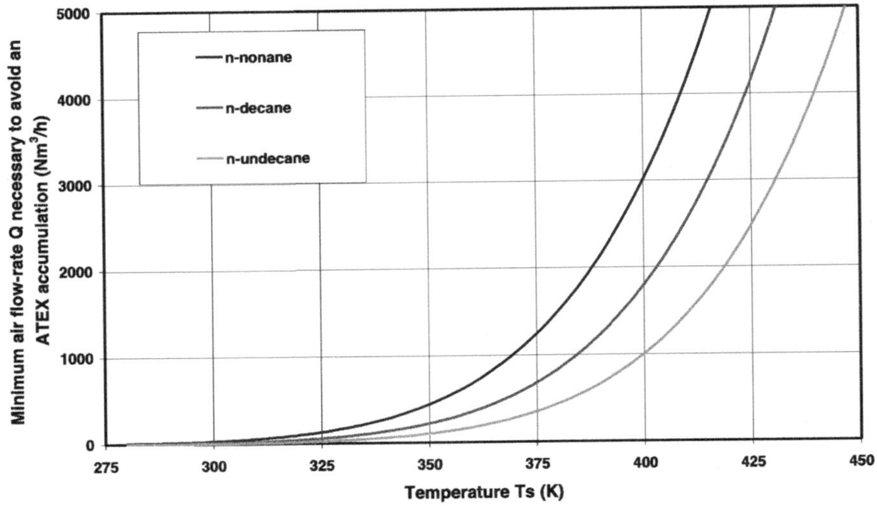

Figure C14.1. *Variation of the minimum air flow to avoid the formation of a significant volume of ATEX, downstream of a puddle of n-nonane, n-decane or n-undecane, as a function of the temperature T_S of the sole (the area of the puddle is assumed to be 20 m^2 and the air velocity above its surface is assumed to be 0.3 m/s). For a color version of this figure, see www.iste.co.uk/chaineaux/mechanisms.zip*

Moreover, it appears that, for an air flow of 4,000 Nm^3/h at the start-up of the kiln, the temperature of the sole would have been 415 K (142°C) at least, for a sufficient volume of ATEX to be formed downstream of a puddle of n-nonane. Similarly, for a puddle of n-decane, the minimum temperature of the sole for a sufficient volume of ATEX to be formed would be 425 K (152°C). Finally, for kerosene whose volatility characteristics are close to those of n-nonane and n-decane, a sufficient volume of ATEX could be formed if the temperature of the sole reached at least 140–150°C.

Finally, it can be supposed that the explosion resulted from the massive vaporization of a puddle of kerosene spread on the sole and from the mixing of this vapor and air.

C14.1.4.2. *ATEX ignition source*

C14.1.4.2.1. Hypothesis of an auto-ignition

To explain the ignition, the hypothesis of an auto-ignition has been considered first, but the ATEX should have been heated up to at least 210°C and maintained at this temperature for a sufficiently long time, corresponding to the induction time of the auto-ignition. Actually, with the vaporization phenomenon being endothermic, the temperature of the sole rather decreased as the jet of kerosene continued to spread on it. The hypothesis of an auto-ignition could thus be ruled out because the temperature was too low.

C14.1.4.2.2. Hypothesis of an ignition by the pilot flame

Since the pilot flame was burning at the time of the explosion, it can be logically supposed that it was at the origin of the ignition of the ATEX. For this hypothesis to be verified, however, ATEX would have been to be present at least up to the height of the kiln axis, i.e. 2.3 m above the sole. This condition has a consequence on the volume of ATEX at the time of ignition, which can be estimated to 100 m^3, a value that is compatible both with the circumstances of its formation and with the effects of the explosion.

C14.1.4.2.3. Summary of the mechanism of explosion

The mechanism of the explosion can be summarized as follows:

– the main burner was fueled with kerosene and operated normally in successive cycles, lasting approximately 1 h 30 in total, to dry the kiln refractory that had just been replaced;

– these sequences of ignition-extinction heated the sole of the kiln to at least 140°C;

– the kiln was then started for a production phase;

– during this phase, the burner had a malfunction (excessive kerosene flow or insufficient atomization) that caused it to shut down, while continuing to produce a jet of liquid kerosene, which spread on the sole;

– in contact with the hot sole, the kerosene vaporized and the vapor was mixed with air to form an ATEX with a sufficient volume to reach the pilot flame which ignited it.

C14.1.4.3. *Influence of the type of fuel used to start the kiln*

It can be asked whether the use of kerosene in place of diesel fuel, which is usually used as a fuel for the burner during the start-up phase, could have favored the occurrence of the explosion:

– diesel fuel is a liquid significantly less volatile than kerosene;

– by assimilating the volatility of the diesel fuel with that of n-undecane, Figure C14.1 shows that a sole temperature of 440 K (167°C) would have been sufficient for diesel fuel to vaporize and form an ATEX able to reach the pilot flame;

– therefore, the conditions of the explosion would have been met at 167°C with the diesel fuel, whereas they were met between 140°C and 150°C for kerosene;

– however, the temperature reached by the sole at the end of the refractory drying operation could easily have exceeded 140°C;

– this difference of 20–30°C is not sufficient to have to consider that the use of kerosene favored the explosion, which would have happened anyway, but a little later.

Lessons from the investigation of an explosion of kerosene vapor in contact with a hot surface

When a flammable liquid is brought into contact with a material at a temperature close to its boiling point, a massive vaporization of the liquid occurs and an ATEX is formed by dilution with air flowing at the surface of the material. Depending on the conditions of air velocity and the extent of the wet surface, the ATEX formed may have a sufficient volume to be ignited by a source of ignition present at some distance from the material.

– In the present case, whether the ratio of kerosene flow rate to air velocity was inadequate or the atomization of kerosene was not of sufficient quality, no combustion was possible (either the burner did not ignite or it flamed out if it was on), although the pilot flame burned constantly; there should have been an automatic shutdown of the kerosene flow, but this device failed;

– The spreading of liquid kerosene on the sole of the kiln, continuously heated as it can be during a drying phase of the refractory, must be considered as a situation with a significant explosion risk and appropriate preventive measures must be defined and implemented strictly, such as the immediate and automatic shutdown of the kerosene flow.

Case 15

Explosion of Volatile Hydrocarbons

C15.1. Case presentation

An explosion occurred in a tank, which was used for the recovery of various flammable liquids, collected in activities of vehicle maintenance and repair.

C15.1.1. *Description of the facility and the circumstances of the explosion*

The facility essentially consisted of:

– a buried tank for the recovery of used oil and other more or less volatile and flammable liquids;

– a device for the recovery of these liquids and their transfer to the tank, through a pipe installed in a pit.

C15.1.2. *Description of explosion damage*

C15.1.2.1. *Human effects*

The operator who was cutting off a steel rod and who was certainly in the immediate vicinity of the inspection shaft above the tank was severely injured and their hair was slightly burned.

C15.1.2.2. *Mechanical effects*

The explosion produced two types of mechanical effects:

– the rupture of mechanical components and the projection of numerous debris and projectiles of widely varying sizes, including:

- the plate closing the manhole of the tank, and some of the broken bolts fixing it to the tank,

- the cap that closed the tank drain hole,

- a little trailer on wheels that was above the inspection shaft, which was overthrown,

- the front wheel of this trailer;

– the generation of an aerial overpressure wave.

C15.1.2.3. *Thermal effects*

The explosion produced only limited thermal effects: on the wall of the pit, near the outlet of the pipe of about 50 mm diameter which was connected to the tank as a breathing vent, the trace of a jet of soot is visible. This trace demonstrates that the flame of the explosion came out of the pipe.

C15.1.3. *Mechanism of explosion*

The manhole and the cap closing the drain hole had to sustain the overpressure developed in the tank. This pressure may have only resulted from the ignition of an ATEX, which was formed in the tank.

To establish the mechanism of the explosion, the following steps have been followed:

– identification of the constituents of the ATEX formed;

– identification of the ignition source;

– estimation of the overpressure developed in the tank and of the combustion energy;

– comparison of the effects of the explosion with those expected during the rupture of the manhole plate, based on the value of the explosion energy released.

C15.1.3.1. *Nature of the ATEX constituents*

Nature of the oxidizer

The constituent oxidizer of ATEX was air, which was naturally introduced into the tank during the last emptying of its liquid contents.

Nature of the fuel

The combustible present in the ATEX was a mixture of the vapors emitted by the used liquids collected in the tank, which were various flammable liquids, the most volatile ones likely to be present being two- or four-stroke engine fuels.

In fact, it can be said that the ATEX could not be made of low volatile liquids or at any rate have a flash point FP higher than the ambient and necessarily contained vapors emitted from a liquid with a flash point significantly lower than the ambient, like automotive fuels.

The minimum concentration of the volatile constituent has been assessed by taking into account:

– the fact that the sky of the tank was a fully enclosed volume, so that a liquid-vapor balance was established between the mixture of vapors contained in the tank, and the mixture of the liquids it contains (the presence of the breathing vent has no effect on this balance);

– this balance is such that the composition of the vapor mixture depends only on the composition and temperature of the liquid mixture.

The diversity of the liquids collected in the tank does not allow us to determine the compositions of the mixture of liquids and the mixture of vapors at the time of the explosion, precisely.

On the contrary, this composition can be estimated by simulation, assuming that the mixture of liquids consisted only of the following two pure liquids, of very different volatility:

– a weakly volatile liquid A, the most abundant;

– a highly volatile liquid B, the least abundant.

The liquid A simulates all of the low volatile components of the mixture, while the liquid B simulates the unknown highly volatile compound.

Principle of the simulation and data

Two liquids A and B were selected based on their flash point, compared to those in Table C15.1:

– the low volatile liquid A is assumed to be tetradecane $C_{14}H_{30}$;

– the very volatile liquid B is assumed to be pentane C_5H_{12}.

Type of component	Molar mass (g/mole)	Flash point FP (°C)	Lower explosion limit (LEL) (% volume)	Saturated vapor pressure of each pure component at 20°C (P_{sat}) (Pa)
Liquid A: $C_{14}H_{30}$	200	100	0.5	$P_{sat}(A) = 1.1$
Liquid B: C_5H_{12}	72	- 49	1.5	$P_{sat}(B) = 58{,}279$

Table C15.1. *Physicochemical characteristics of liquids A and B*

To conduct the simulation, the following characteristics must be taken into account:

– x_A the molar fraction of A in the A–B mixture;

– x_B the molar fraction of B in the A–B mixture;

– x'_A the mass fraction of A in the A–B mixture;

– x'_B the mass fraction of B in the A–B mixture;

– M_A the molar mass of A, i.e. 72 g/mole;

– M_B the molar mass of B, i.e. 200 g/mole;

– $p_{sat}(A)$ the saturating vapor pressure of pure A at 20°C;

– $p_{sat}(B)$ the saturating vapor pressure of pure B at 20°C;

– $p_{sat}(A)_{mix}$ the partial pressure of A at 20°C in the tank, assumed to be in equilibrium with the A–B mixture;

– $p_{sat}(B)_{mix}$ the partial pressure of B at 20°C in the tank, assumed to be in equilibrium with the A–B mixture;

– atmospheric pressure $P_{atm} = 101{,}300$ Pa.

The A–B mixture is supposed to be an ideal solution governed by the Raoult law, which is written (relationship [C15.1]):

$$p_{sat}(A)mix = x_A \text{ x } p_{sat}(A) \qquad [C15.1]$$

$$p_{sat}(B)mix = x_B \text{ x } p_{sat}(B)$$

n_A being the number moles of A in the mixture A-B and n_B being the number of moles of B in the mixture A-B, the equation [C15.2] can be written:

$$x_B = n_B/(n_A + n_B) \qquad [C15.2]$$

and, since $n_A >> n_B$, the relationship becomes [C15.3]:

$$x_B = n_B/n_A \qquad [C15.3]$$

The purpose of the simulation is to determine for which value of the mass fraction of B, x'_B, an ATEX is present in the tank:

– at 20°C, the ratio of the saturated vapor pressures of A and B is:

$$P_{sat}(A)/P_{sat}(B) = 1.1/58{,}279 = 0.00002$$

– the difference in the volatility of liquids A and B is such that vapor of B is predominant in the tank, the vapor of A being almost absent;

– the lower explosion limit (LEL) of the mixture contained in the tank is therefore equal to that of pure B or 1.5% volume;

– the LEL will be reached if the saturated vapor pressure of B in the tank is such that:

$$p_{sat}(B)_{mix} > 0.015 \text{ x } P_{atm} = 1{,}519 \text{ Pa}$$

– the Raoult law gives:

$$x_B = p_{sat}(B)_{mix}/P_{sat}(B)$$

– the molar fraction of B in the liquid must therefore be such that:

$$x_B > 1{,}519/58{,}279 = 2.6\% \text{ volume.}$$

Simulation results

Finally, the minimum estimated value of x'_B is:

$$x'_B > 2.6 \text{ x } (72/200) = 0.9\% \text{ weight}$$

Basically, a pentane ATEX is formed at 20°C in a tank containing a pentane–tetradecane mixture, if the pentane mass fraction in the mixture is at least 1%.

Strictly speaking, the nature of the low-volatile liquid A is irrelevant since it is assumed to be much less volatile than liquid B, and therefore the sky of the tank contains practically only B vapor.

C15.1.3.2. *Nature of ignition source*

The explosion occurred when the operator cut off half the rod, and it means that ATEX contained in the tank has been ignited during the cutting off operation, according to the following mechanism:

– such a cutting off operation produces a lot of friction sparks which are made up of small incandescent particles;

– these particles are brought to a very high temperature and constitute a certain source of ignition of an ATEX;

– in this case, the operator's arrangement with respect to the pit was such that a spark was thrown into the used liquid recovery bin. It passed through one of the holes in the plate covering the bin and it reached the bottom of the bin, where the ATEX was present;

– the combustion of the ATEX was initiated and the flame propagated to the tank, going up into the pipe connected to it.

C15.1.3.3. *Estimation of the overpressure developed in the tank*

From the estimated composition of the ATEX, it is possible to estimate:

– the maximum explosion pressure;

– the combustion energy.

When an ATEX is contained under atmospheric pressure in a closed container of sufficient mechanical strength, its explosion develops a pressure in the container with the following characteristics:

– if the ATEX concentration is optimal, that is, close to the stoichiometric concentration, the developed pressure is maximum: it is called the "maximum explosion pressure" or P_{max} and, expressed in absolute bar, it is close to 9 bar (see section 1.1.4.3.1);

– at LEL, the developed overpressure is about 3 to 4 bar;

– the result is that the overpressure developed for any concentration between LEL and stoichiometric is between 3 bar and P_{max}.

In the present case, the LEL is assumed to be 1.5% volume and the stoichiometric concentration is around double the LEL, i.e. 3% volume.

When the container does not have sufficient mechanical strength, a rupture of its most fragile element occurs when the pressure developed by the explosion reaches its rupture pressure.

In the present case, we know that the plate securing the manhole was torn off.

The fracture pressure P_r of this plate can be estimated from its dimensional and mechanical strength characteristics:

– its surface area of 0.28 m^2;

– the diameter (5 mm) and number of bolts (16) securing it to the dip and either sheared or punched into the plate;

– the estimated fracture stress of the steel, $\sigma_r = 400$ MPa.

The calculation leads to a value of $P_r = 4.5$ bar, which is consistent with the range of the overpressure values developed by the explosion, between 3 and 8 bar.

However, this value, which is independent of the volume of the tank, presents a significant uncertainty, and it is not possible to determine the concentration of ATEX between LEL and stoichiometric composition.

It should be added that, in addition to the tearing of the manhole plate, the explosion may have broken a certain weld in the cylindrical part of the tank which may have become leaky.

C15.1.4. *Estimation of energy released by explosion*

The explosion resulted from the release of an energy E_{lib} equal to the combustion energy of the ATEX.

This energy depends on the following factors:

– the volume of ATEX;

– the composition of ATEX.

Assuming a volume of ATEX equal to 2 m^3, and based on its concentration, i.e. between 1.5% and 3% volume and on the combustion heat of the pentane, i.e. 44.6 MJ/kg, the explosion released an energy between 3.5 and 7 MJ.

C15.1.4.1. *Estimation of the energy E_{eff} involved in the production of damage*

The energy required to produce the explosion damage is directly related to the energy released by the combustion.

However, it must be considered that any explosion is characterized by its yield ρ, and therefore that the damage was produced by an energy that represented only a fraction of the energy released by the combustion of ATEX.

As indicated in section 15.1.2, the mechanical damage of the explosion resulted in the projection of numerous debris and an aerial pressure wave.

Each of these effects resulted from the production of hot burned gases inside the tank, and which expanded violently outside, following the tearing of the manhole plate.

The order of magnitude of the energy E_{eff} released during this expansion can be estimated at 2 MJ, based on the volume of gases inside the tank, estimated at 2 m^3, and the rupture pressure of the manhole plate, estimated at 4.5 bar.

This energy value corresponds to an explosion having a TNT equivalent of about 450 g.

This value is consistent with that of the energy released by the combustion of ATEX, since it is slightly lower, which reflects the existence of a yield ρ less than 1 in the transfer of energy between the combustion of ATEX and the expansion of explosion gases.

C15.1.4.2. *Conclusion about the mechanism of explosion*

Ultimately, the search for the mechanism of explosion led us to the following conclusions:

– at the time of the explosion, the vessel contained vapors emitted by a flammable liquid that was significantly more volatile than the used liquids normally introduced into the vessel;

– it is possible to say that it is a liquid with a flash point much lower than the ambient temperature, such as a premium gasoline; according to this hypothesis, the calculation shows that it would be sufficient for the contents of the tank to have a premium gasoline mass concentration of only 1% for an ATEX to be present in the tank;

– the ATEX diffused inside the pipe that connected it to the recovery bin until it reached the base of the bin;

– the ATEX was likely ignited by a friction spark produced during the rod cutting operation by an operator in the vicinity of the pit, which fell into the recovery bin;

– the explosion caused the bolt-on manhole plate to be pulled out and projected out of the tank (as well as numerous debris to be projected), allowing the overpressure developed in the tank to be estimated at several bars;

– the value of the energy released by the explosion can be estimated at 2 MJ, which corresponds to an ATEX volume of approximately 2 m^3 and a flammable vapor concentration well above the LEL, which is estimated at 1.5% volume.

Lessons to be learned from the investigation of explosion of volatile hydrocarbons

The ATEX that exploded in the tank had been formed from a mixture of vapors emitted into the sky of the tank, by a mixture of the following flammable liquids:

– at least one liquid with pentane-like volatility (FP = – 49°C);

– very low volatile liquids with flash points well above ambient temperature.

An ATEX can be formed as soon as the mass fraction of the volatile constituent in the mixture is sufficient, i.e. about 1%.

Case 16

Explosion in a Spray Dryer of Powdered Milk

C16.1. Case presentation

The mechanism of an explosion that occurred in a spray dryer producing powdered milk had to be determined.

C16.1.1. *Description of the facility*

The main elements of the spray dryer are shown in Figure C16.1.

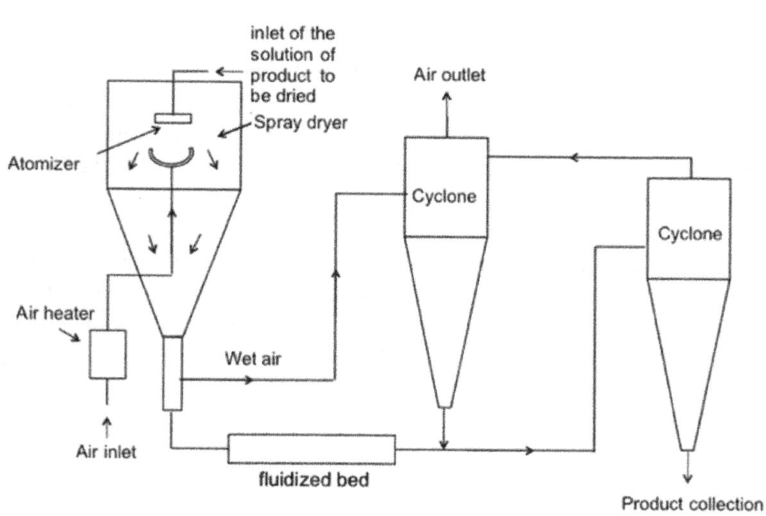

Figure C16.1. *Scheme of the spray dryer*

In the present case, the main parts of the spray dryer and their characteristics were the following:

– the drying air heating unit: this unit produced a flow of drying air at a rate of 100,000 m^3/h, which was heated up to 195°C by an air stream burner in a heat exchanger, up to the set temperature, i.e. 215°C;

– the spray dryer itself: the evaporating milk droplets were produced at the top of the dryer, by atomization of a solution by the nozzle of a turbine, and they were entrained into the downward flow of the drying air (co-current dryer);

– the drying air dust removal equipment, where the dust and wet air came out of the dryer at a temperature close to 100°C and was directed to a cyclone;

– the milk powder collection equipment: at the bottom of the dryer, the milk powder was introduced at 60°C into a vibrating fluidized bed, the fluidization air of which was at room temperature; the dry powder came out of the bed at a rate of 5 T/h, for a moisture content of around 4%, while the fluidization air was dedusted in a cyclone.

C16.1.2. *Circumstances and effects of the explosion*

The explosion occurred several minutes after a failure of the heat exchanger's steam supply resulting in less heating of the drying air, which was automatically offset by an increase in its heating by the burner. However, according to the operator's experience, this method of heating the drying air made it more difficult to regulate its temperature to the set value of 215°C.

The dryer was not damaged by the explosion as it was protected by three well-sized explosion vents that normally opened. Similarly, the fire extinguisher system fitted to the dryer functioned normally, and the fire-extinguishing water, introduced into the dryer and the fluidized bed, quickly extinguished the fire that had been initiated by the flame of the explosion.

This disaster had been preceded one year earlier by another in very similar conditions, except that the fire initiated by the explosion at the bottom of the dryer had not been stopped fast enough, because the water spray nozzles of the fluidized bed were under the level of the milk powder which rendered them inoperative. The fire then propagated to the cyclone, downstream of the bed.

C16.1.3. *Flammability and explosion characteristics of milk powder*

It is necessary to know certain flammability characteristics of milk powder, not only in order to investigate a disaster involving this product, but also to define the means of prevention and protection to be installed to prevent the occurrence of a new disaster.

NOTE.– For any powdery product, the values of these characteristics are only accessible by experimentation, and they are available in the literature for the most common products. It should be noted, however, that some of them depend heavily on the size of the product and its moisture content. Therefore, they must be considered as a range of values.

C16.1.3.1. *Lower explosion limit (LEL)*

Also referred as the "minimum explosive concentration", the LEL at 20°C of the milk powder dispersed in air at ambient temperature is worth 60 g/m^3, and this value becomes 30–40 g/m^3 when the air-powder mixture is at 100°C.

C16.1.3.2. *Minimum oxidation temperature*

The relatively low oxidation temperature of powdered milk, between 160 and 290°C, makes it a very easily oxidizable product.

C16.1.3.3. *Auto-Ignition temperature of an air suspension (cloud)*

The minimum ignition temperature of an air suspension of powdered milk is worth between 340 and 600°C.

C16.1.3.4. *Minimum ignition temperature of a layer*

The minimum ignition temperature of a deposit (a layer) of powdered milk depends on the thickness of the layer, the surrounding atmosphere being at room temperature.

In a hot environment, as is the case in the airflow of a spray dryer, the relationship between the critical temperature of self-heating, which leads to the ignition of the layer, and the thickness of the layer can be established experimentally. Depending on whether it is a milk powder of minimum or maximum oxidizability, the curves in Figure C16.2 are used to determine the minimum temperature of the atmosphere, for which a layer of this product of a given thickness will give a self-heating leading to its ignition (Ineris DRA 35, report Omega 11 2005).

Similarly, these curves are used to determine the minimum thickness of a layer of powdery product, for which a given temperature will initiate its self-heating leading to its ignition.

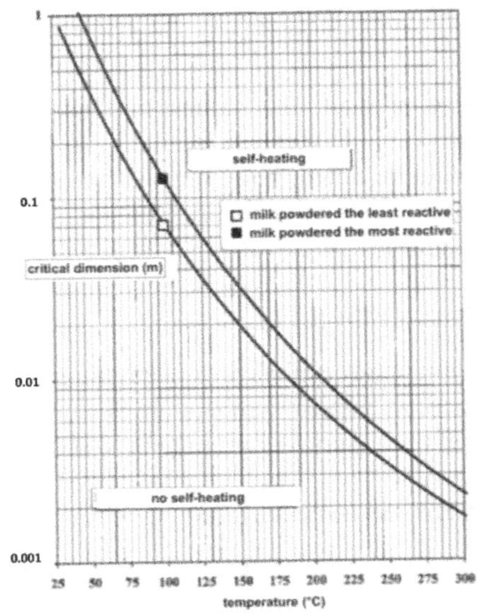

Figure C16.2. *Variation of the temperature and thickness thresholds of a milk powder layer beyond which its self-heating will occur and lead to its self-ignition*

C16.1.3.5. *Minimum ignition energy (MIE)*

The minimum ignition energy of an air suspension of powdered milk is between 30 mJ and 1 J, making it a product that is few sensitive to ignition by electrostatic phenomena.

C16.1.3.6. *Maximum explosion pressure (P_{ex})*

The pressure variation curve developed in a 20-dm^3 sphere during the explosion of an ATEX consisting of an air suspension of powdered milk taken under normal temperature and pressure conditions, is similar to the curve in Figure 1.5 relative to an air–methane ATEX. In the case of milk powder, the values of the maximum explosion pressure P_{ex} and the maximum rate of pressure rise K_{st} are the following ones:

$P_{ex} = 6.7 + 1 = 7.7$ bar abs;

$K_{st} = 72$ bar. m/s.

C16.1.4. *Mechanism of explosion*

For an explosion to occur in the dryer, the following two conditions had to be met:

– the presence of an ATEX in the dryer at least locally;

– the presence of an ignition source in this ATEX.

C16.1.4.1. *Presence of an ATEX in the dryer*

On the basis of the values of the hot airflow rate and the powdered milk flow rate (100,000 Nm^3/h and 5 T/h, respectively) for the nominal operating conditions of the dryer, it appears that the average concentration of the air powdered milk suspension was close to 50 g/m^3, which is slightly higher than the LEL at 100°C.

In addition, the atmosphere at the bottom of the dryer contains water vapor from the evaporation of the water contained in the milk droplets. It is difficult to quantify the water vapor content of this atmosphere and thus also its influence on the flammability of this atmosphere. However, it is possible to say that, qualitatively, its influence decreases its flammability in relation to that of a dry atmosphere. In any case, it can be admitted that, because of an abnormally high drying air temperature at the top of the dryer, the atmosphere at the bottom of the dryer was an ATEX.

C16.1.4.2. *Ignition source in ATEX*

At the time of the first accident, a combustion zone had been discovered at the wet air extraction pipe located on the dryer axis. This outbreak had resulted from the self-ignition of a layer of powdered milk that had accumulated between this horizontal cylinder-shaped pipe and a jumper-shaped piece that had been welded to the top of the pipe tip, precisely to prevent a layer from accumulating. The arrangement of this piece in relation to the pipe allowed the thickness of the layer to exceed the limit value between 7 and 15 cm, drawn from the curves in Figure C16.2.

The same mechanism is the same for the second disaster:

– the abnormally high temperature allowed both the presence of an ATEX at the bottom of the dryer and the self-heating of a layer that appeared on the extraction pipe of the wet air;

– the self-ignition of the layer;

– the ignition of ATEX, either by direct contact with the burning layer or by an incandescent particle falling from the air extraction pipe of the dryer.

C16.1.4.3. *Explosion event tree*

From the information collected during the investigation, the events tree has been built; it enables us to visualize each of the elementary causes that constitute the global mechanism of the explosion.

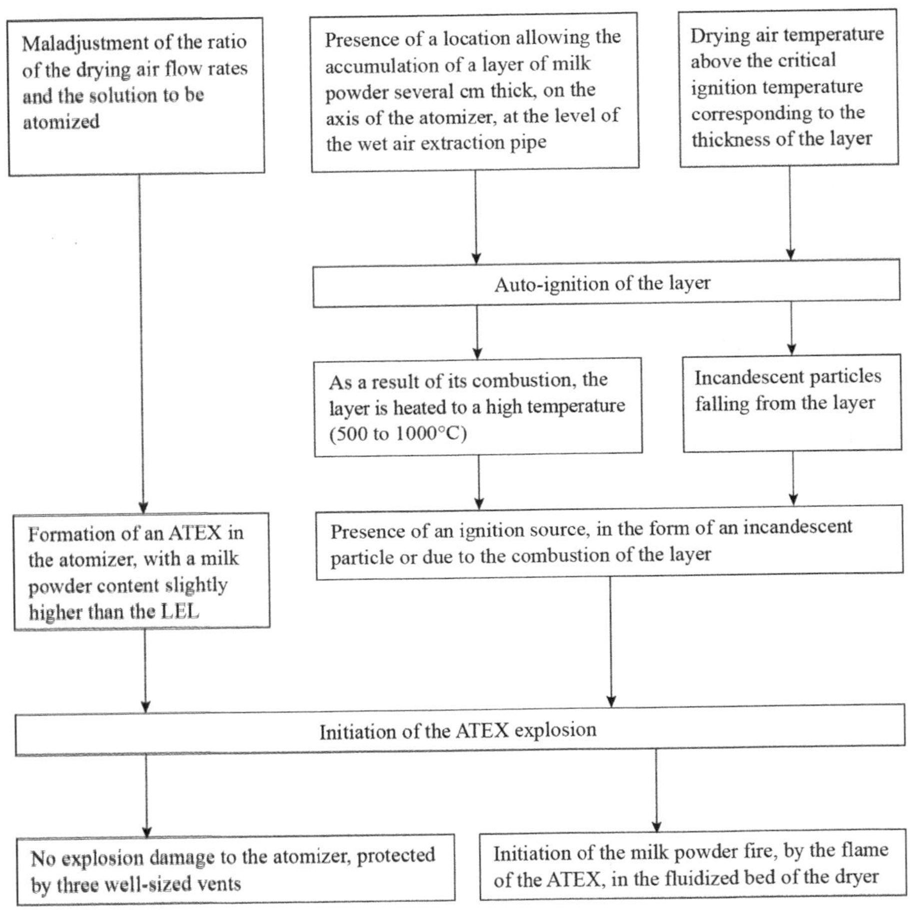

Figure C16.3. *Events tree showing each of the elementary causes of the explosion*

Lessons learned from the expertise of an explosion in a dryer

Explosion and fire are important risks in the operation of a process of drying a powdery combustible product in a spray dryer and in its ancillary equipment for the dedusting of the dried product:

– under normal operating conditions, an ATEX may form because the composition of the atmosphere in the dryer is not very far from the lower flammability limit (LEL);

– among the possible ignition sources of this ATEX, it is necessary to prevent those that could result from the self-ignition of the accumulated product in the form of a layer in a hot environment.

It is therefore essential that the installations are properly protected, both against the risk of explosion and the risk of fire:

– the prevention of ignition risks must be ensured by rigorous temperature control in each part of the installation and by preventing the formation of layers or by limiting their thickness;

– this control must be based on the knowledge of the thresholds beyond which the phenomenon of self-ignition of a product layer occurs:

- the maximum ambient temperature corresponding to a given thickness,

- the maximum thickness of a layer corresponding to a given ambient temperature,

- the dryer and the dust removal equipment must be protected by an explosion vent or other means of controlling the overpressure effects of the explosion,

- the dryer must also be equipped with automatic fire extinguishers in order to quickly control any fire outbreak,

- finally, it is useful to monitor the CO content of the air coming out of the dryer, since any significant presence of this gas, absent from the ambient air, is a sign of combustion and therefore a sign of a start of fire;

– if an explosion occurs, it is common for it to initiate a fire, but an explosion can also be initiated by a layer of powdered product, the self-heating of which has resulted in self-ignition.

Case 17

Explosion in a Wood Waste Grinding Facility

C17.1. Case presentation

An explosion occurred in a grinding facility producing pellets from wood waste in the form of woodchips (see Figure C17.1).

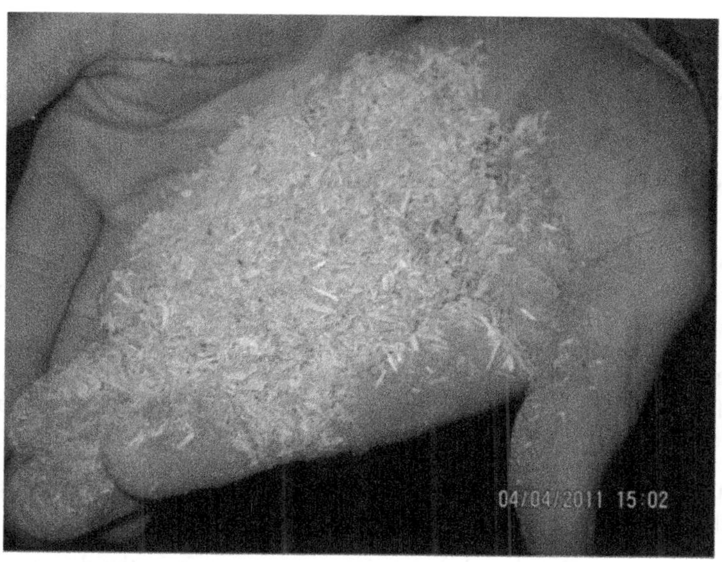

Figure C17.1. *Aspect of the dry ground woodchips*

For a color version of all figures, in this case study, see www.iste.co.uk/chaineaux/mechanisms.zip.

The combustible nature of wood dust and the absence of any other element that could cause the explosion suggest that it resulted from the ignition of an ATEX that was formed locally in the facility.

The mechanism of the explosion had to be determined, particularly the identification of the location where the ATEX was present in the facility and the nature of the ignition source.

C17.1.1. *Description of the facility*

From upstream to downstream, the production line consisted of the following equipment:

– a first grinder, the so-called "wet grinder", insuring the coarse grinding of the woodchips;

– at the outlet of the wet grinder, a bucket conveyor for the transport of the wet ground woodchips;

– a dryer consisting of three-layered slow-moving conveyors with metallic belts, on which the wet ground woodchips were flowed through by an upward hot air stream;

– at the outlet of the dryer, a bucket conveyor carrying the dried ground woodchips to a hopper, feeding a second hammer grinder, the so-called "dry grinder", operating at room temperature;

– under the dry grinder, a set consisting of:

- a casing collecting the ground woodchips, and maintaining a slight depression relative to atmosphere, so as to facilitate their collection in the casing (the depression was obtained by means of an extraction fan, flowing air through a cyclone separating wood dust from air),

- a rotary lock valve located at the base of the depression casing and allowing the transfer of ground woodchips downstream, while keeping the depression in the casing;

– a bucket conveyor carrying the ground woodchips between the outlet of the rotary valve and the pelletizer hopper;

– a pelletizer transforming the particles coming out from the dry grinder after having been slightly re-moistened into pellets;

– finally, a bucket conveyor carrying the pellets from the pelletizer to a bagging unit.

C17.1.2. *Circumstances of the explosion*

At the time of the explosion, the production conditions of the pellets were normal.

However, the high degree of wear of the belts of the dry grinder should be considered as a special circumstance of the explosion:

– as the facility was being operated, these belts showed rapid wear;

– this wear provoked the rupture of several mesh threads at the edge of the belt, as well as the rupture of the needles holding the mesh;

– after the explosion, metal particles of various shapes and sizes were found in the dust downstream of the dryer, as well as in the input of the pelletizer (see Figure C17.2).

Figure C17.2. *Metal particles found downstream of the dryer (the square side of the paper grid is 5 mm)*

C17.1.3. *Effects of the explosion*

C17.1.3.1. *Mechanical effects*

The explosion produced the deformations of certain elements located downstream of the grinder:

– the bulging of the parallelepiped fitting between the decompression casing and the lock valve;

– the bending of the screw feeding the lock valve.

C17.1.3.2. *Thermal effects*

The explosion was followed by a fire, that first developed in the buffer hopper and then propagated to the dry grinder's feed bucket conveyor, up to the dryer, which has been severely damaged.

C17.1.4. *Flammability characteristics of the wood dust*

The material implemented in the dry grinder is a mixture of wood particles that are fibrous and have more or less elongated shapes.

It is difficult to specify the size distribution of the particles mixture, but it is possible to say that the proportion of particles with an average size of less than 0.3 mm is significant (see Figure C17.1).

The flammability characteristics of wood dust depend on many factors such as the type of wood, the size of the wood particles and their moisture content. These are as follows (the values indicated should be considered as orders of magnitude):

– minimum ignition temperature of a cloud: this is the lowest temperature for which a suspension of dust in the air ignites spontaneously, with no energy input other than its own temperature; for a softwood, it is close to 500°C;

– minimum ignition temperature of a dust layer: this is the lowest temperature for which a dust layer ignites spontaneously in the presence of air, without energy input other than the temperature of the medium on which the layer is located. This characteristic depends on the thickness of the layer, and for a 5-mm thick layer of softwood dust, it is close to 300°C;

– minimum ignition energy (MIE): this is the lowest energy that must be brought locally to a dust suspension in air to get its ignition; in the case of wood dust, the value to be retained is about 20 mJ;

– lower explosion limit (LEL): this is the lowest concentration of a dust dispersed in air for which a flame still propagates; for a wood dust, it is close to 40 g/m^3.

C17.1.5. *Determination of the mechanism of explosion*

Based on the nature and location of the effects, the explosion was initiated in the dry grinder, with an ignition source that remains to be identified.

C17.1.5.1. *Possible ignition sources*

There is no source of ignition present in normal operation in the grinder. However, different sources of ignition may occasionally occur under abnormal operating conditions. The four following phenomena may a priori be at the origin of the ignition:

– electrostatic spark;

– auto-ignition phenomenon of a suspension of wood dust in air;

– self-heating of a wood dust layer;

– mechanical friction spark.

C17.1.5.2. *Examination of each possible source*

C17.1.5.2.1. Electrostatic spark

For a phenomenon of electrostatic discharge (or electrostatic spark) to occur, it must be preceded by a phenomenon of generation and accumulation of electrostatic charges (friction, separation of bodies of different nature, etc.) during which the electrostatic voltage between two bodies increases strongly; it is when this voltage reaches a threshold called "breakdown voltage" (several kilovolts) that a discharge occurs in the form of an electrostatic spark between the two bodies, and the energy released in this discharge may be sufficient to ignite an ATEX.

The accumulation of a sufficient quantity of electrostatic charges within a dust is only possible if the dust is sufficiently resistive, i.e. it prevents the flow and then the neutralization of electrostatic charges.

Because of its moisture content, the wood dust is not sufficiently resistive, so that the accumulated charges are neutralized as they appear and the electrical voltage of the dust suspension never reaches the breakdown voltage. Then, it is not possible that an electrostatic phenomenon could be the ignition source.

C17.1.5.2.2. Auto-ignition of a suspension of wood dust in air

The grinding was done at room temperature and, in normal operation, the grinder had no hot surface. However, the following abnormal operating situation should be considered:

– a bearing could have bound and heated intensely, and it must therefore be considered that, near the bearing, the temperature of the internal wall could reach locally the minimum auto-ignition temperature of the cloud of wood dust;

– however, the distance of the grinder bearings from its wall limits the surface that would be raised to high temperature in case of binding.

For an auto-ignition to occur, the ATEX should remain in contact with the hot wall for a sufficiently long induction time. However, the ATEX formed in the grinder does not meet this condition since it is very turbulent. Therefore, the possibility of an auto-ignition must be left out.

C17.1.5.2.3. Self-heating of a wood dust layer

When a wood dust is deposited on a surface, it may give rise, under certain conditions of ambient temperature and thickness of the layer, to a self-heating phenomenon. This combustion first occurs slowly, and then it accelerates and becomes a hot combustion with a temperature of several hundred degrees so that, if the deposit is lifted as a cloud into the air, the burning dust grains are heated to a temperature sufficient to ignite the formed ATEX.

From the ambient temperature, self-heating of the wood dust is only possible for a deposit of several centimeters thick. However, the operating conditions of the grinder (parts in rapid rotation, no dead volume) are such that no deposit of this thickness can be formed.

The self-heating of a wood dust layer cannot be the cause of the ignition.

C17.1.5.2.4. Mechanical friction spark

The impact or friction of two bodies on each other may, depending on their nature and hardness, give rise to a friction spark of sufficient energy to initiate the ignition of an ATEX.

It has been observed that metal particles from the metal belts inside the dryer were introduced into the grinder at the same time as the wood (see Figure C17.2).

It must be considered that the shock of one of these particles caused could produce a friction spark having an energy at least equal to the MIE of the wood dust.

In any case, it must be considered that it is a mechanical friction phenomenon that is the most probable cause of the ignition.

Lessons learned from the investigation

The installations in which processes requiring to implement combustible dusts have a high risk of explosion and fire: even though these processes use particles of high particle size, fine particles are also inevitably present.

It is therefore essential to carry out a thorough and rigorous analysis of the risks of explosion and fire in all parts of the facility, in order to define the means of equipping them, to prevent explosions and to effectively protect against them, if an explosion does occur.

Several ignition sources are possible:

– electrostatic spark;

– auto-ignition phenomenon of a suspension of dust in air;

– self-heating of a dust layer;

– mechanical friction spark.

It is common that the explosion of an ATEX, consisting of a mixture of fine particles suspended in air, initiates the fire of material with high particle size and that the fire propagates at great distances downstream and/or upstream and produces considerable damage, in addition to that produced by the initial explosion.

Alternatively, a fire may be initiated by the flame of an ATEX or be a result of the self-heating of a layer.

In processes involving the transport of dust in an air stream, the risk of ATEX ignition by electrostatic sparks can be particularly acute for the following reasons:

– the concentration of this stream is variable, depending on the process: it is important in pneumatic transport (in principle the dust concentration is much higher than the UEL) and it is low in a dedusting circuit (in principle lower or close to the LEL), but it is not excluded that, in the event of changes in operating conditions, an ATEX is formed locally;

– in all cases, the airflow rate should be sufficient to prevent the dust from settling in the pipes (typically at least 20 m/s);

– the friction between the dust grains and the walls of the transport line generates electrostatic charges;

– these charges accumulate on the grains if the resistivity of the dust is sufficiently high (this is the case of organic products, especially when they are dry);

– the energy is stored on these grains and can be released in an electrostatic spark;

– an ignition may occur if the energy released is higher than the minimum ignition energy (MIE) of the dust;

– it is therefore essential to protect the installations against the risk of ignition of an ATEX by an electrostatic source, by appropriate prevention measures (equipotential connections between equipment, use of antistatic filter bags, work clothes and shoes antistatic and non-resistive soils).

Independently of the electrostatic risk, the dust can also deposit as a layer in an elbow of the de-dusting pipe and give rise to a self-heating phenomenon leading to a fire.

Case 18

Explosion of a Chloroduct

C18.1. Case presentation

A chloroduct is a long steel pipe used for the transport of gaseous chlorine.

The explosion of a chloroduct (3.6 km length, 219 mm diameter), which was electrically heated on its outer surface and which had several expansion lyres, had to be investigated.

C18.2. Circumstances of the explosion

With a view to a planned shutdown of the installations, the chloroduct had been the subject of a procedure which consisted of closing the isolation valves implanted at its ends, depressurizing it from 6 to 1.2 absolute bars and de-energizing its electrical tracing.

The explosion occurred approximately 25 hours after this shutdown.

C18.3. Effects of explosion

The explosion caused the chloroduct to burst, over a total length of about 200 m, located just upstream its downstream valve (see Figure C18.1): in fact, over this total length of 200 m, the bursting occurred in alternating splintered sections (three sections with a total length of 64 m), separated by apparently intact sections or possibly with bulges (see Figure C18.2).

For a color version of all figures, in this case study, see www.iste.co.uk/chaineaux/mechanisms.zip.

Figure C18.1. *View of the exploded zone, 200 m upstream of its point of entry to the site of use*

Figure C18.2. *View of an unexploded section of chloroduct with bulges*

The burst produced an aerial overpressure wave whose peak pressure can be estimated between 20 and 30 mbar at 80 m from the center of the explosion, based on its effects on the only nearby building.

C18.4. Determination of the explosion mechanism

C18.4.1. *Explosive system identification*

The burst was the result of an internal explosion, which developed a sufficient pressure to locally reach the chloroduct rupture pressure.

In the absence of air in the chloroduct as well as in the absence of any physical phenomenon able to generate an internal overpressure, it can be stated that the explosive system was type 2, which corresponds to an exothermic chemical reaction between two or more reagents or to the decomposition of a single reagent.

The nature of the chemical reagents present in the chloroduct, as well as the causes of the explosion initiation, has been identified from its circumstances:

– during the shutdown of the chloroduct, two simultaneous failures led to the voltage holding of its electrical heating on its downstream part and to the exceeding of the set point, i.e. 60°C; then, the temperature of the hottest point of its inner wall has exceeded 100°C for approximately 24 hours;

– to protect the chloroduct against corrosion, the chlorine must be sufficiently dry, and it has been controlled that the water vapor content of the chlorine did not exceed 20 ppm. Indeed, for such a content, the temperature at which the chlorine reacts exothermically with the steel must be at least 120°C. If this content is higher, a passivation layer deposits on the inner wall of the chloroduct according to the reaction [C18.1]:

$$2\ Fe\ (solid) + 3\ Cl_2\ (gas) = 2\ FeCl_3\ (solid) \qquad [C18.1]$$

– it is likely that water vapor entered the chloroduct through a port maintained open in error for several days near the site of use, and that water combined progressively with $FeCl_3$ to form hydrates ($FeCl_3$, n H_2O), with n = 2; 2.5; 3.5 or 6;

– an approximately 0.7 mm thick layer had been observed on the inner wall of the chloroduct between its point of entry to the site of use and the first upstream expansion lyre; this layer was characterized as trichloride iron hexahydrate ($FeCl_3.6H_2O$);

– this trichloride has a relatively low melting temperature, i.e. 37°C (Gmelin and Meyer 1930) and results in the formation of a $FeCl_3$ solution in the crystallization water according to the reaction [C18.2]:

$$Fe\ Cl_3.6H_2O\ (solid) = FeCl_3 + 6H_2O\ (liquid) \quad [C18.2]$$

– it is established that the liquid phase resulting from the fusion of ($FeCl_3.6H_2O$) is highly corrosive and rapidly attacks steel with intermediate hydrochloric acid formation and release of hydrogen in the following reactions [C18.3] and [C18.4]:

$$FeCl_3 + 3\ H_2O = Fe\ (OH)_3 + 3\ HCl \quad [C18.3]$$

$$Fe + 2\ HCl = Fe\ Cl_2 + H_2 \quad [C18.4]$$

– under these conditions, an atmosphere consisting of a hydrogen–chlorine mixture was formed in the chloroduct;

– the flammability and explosive characteristics of hydrogen–chlorine mixtures are summarized in Table C18.1.

Flammability and detonability characteristics of hydrogen–chlorine mixtures	
Flammability range (Sagim 1996)	LEL = 3.5% volume; UEL: 89% volume
Stoichiometric composition	50% volume H_2 / 50% volume Cl_2
Auto-ignition temperature (A. Zagulin 1933)	AIT = 207°C (at ambient P for a 30% volume H_2 mixture)
Maximum explosion pressure P_{max} generated by the deflagration of a stoichiometric mixture at given P, T conditions and in constant volume	P_{max} = 15 bar (at P_0 = 1.4 bar abs. T_0 = 90°C)
Detonability range (Sagim 1996)	Lower Detonability Limit: LDL = 17.5% volume Upper Detonability Limit UDL = 83% volume
Detonation pressure P_{det} for a stoichiometric mixture	P_{det} = 30.7 bar (at P_0 = 1.4 bar abs. and T_0 = 90°C)
Detonation velocity C_{det} for a stoichiometric mixture	C_{det} = 2,018 m/s
Length of the detonation cell λ at P, T ambient	lambda ≅ 1 mm for a stoichiometric mixture lambda ≅ 600 mm near the LDL

Table C18.1. *Flammability and detonability characteristics of hydrogen–chlorine mixtures*

The mass of the layer of ($FeCl_3.6H_2O$) deposited on the internal wall of the chloroduct, as well as the section of the chloroduct, about 4 dm^2, and the maximum pressure reached during the explosion enables us to estimate at 20% volume the H_2 content of the hydrogen–chlorine mixture and thus infer that it was an atmosphere likely to detonate.

If the mixture did detonate, the detonation was preceded by a phase of deflagration, until the flame propagates for a length equal to the "pre-detonation distance".

To accept the hypothesis of a detonation of a hydrogen–chlorine mixture formed in the chloroduct, different conditions must be fulfilled:

– the cause of the initiation of the hydrogen–chlorine explosion in the regime of deflagration must be identified;

– the cause of the transition from deflagration to detonation must be identified;

– the correspondence between the pressure effects produced by the detonation and the chloroduct rupture pressure must be determined.

Based on literature, the AIT of a hydrogen–chlorine mixture with 30% volume hydrogen is worth 207°C, which does not enable us a priori to explain the auto-ignition of a hydrogen–chlorine mixture with 20% volume hydrogen, at a temperature not much exceeding 100°C.

However, it must be taken into account that the time between the isolation of the chloroduct and the explosion was approximately 25 hours.

The literature reports the results of Russian work (Frolov 1977) on the variation, as a function of the AIT, of the ignition delay Δt of hydrogen–chlorine mixtures, for hydrogen contents between 8 and 18% volume:

– this variation reflects the fact that Δt is inversely proportional to AIT;

– in addition, the influence of hydrogen content on Δt is linear over the interval studied.

An application of these results to the explosion conditions, for $\Delta t = 25$ h, provides the AIT value, i.e. 74°C.

The fact that this value is significantly lower than the maximum temperature reached by the chloroduct means that the explosion of the hydrogen–chlorine mixture could have been initiated by an auto-ignition mechanism, in the chloroduct section without thermal regulation.

C18.4.2. *Estimation of the rupture pressure P_r of the chloroduct*

The rupture pressure P_r of the chloroduct is obtained from the Lamé relation (see section 1.3.2):

$$P_r = 2\ e\ \sigma_r/D$$

where:

– e is the chloroduct thickness or e = 5.6 mm;

– D is the diameter of the chloroduct, D = 219 mm;

– σ_r is the rupture stress of the chloroduct steel (grade 37A), i.e. 370 MPa.

The calculation gives: P_r = 192 bar.

C18.4.3. *Different arguments for a detonation of the hydrogen–chlorine mixture*

The maximum P_{max} deflagration pressure of a hydrogen–chlorine stoichiometric mixture, i.e. 15 bar, is nearly 13 times lower than P_r, and we can therefore exclude that the chloroduct rupture resulted from an explosion propagating in the deflagration regime, even though it is established that the elongated geometry of the chloroduct favored the acceleration of the flame and the development of a pressure locally higher than P_{max}.

Another reason for excluding the hypothesis of a hydrogen–chlorine mixture deflagration is due to the fact that the cross-section of the chloroduct fragments do not exhibit a brittle fracture surface with a chevron structure.

However, it is possible that the detonation of the hydrogen–chlorine mixture locally developed a pressure well above the P_{det} value of 30.7 bar for a stoichiometric mixture (see Table C18.1). According to this hypothesis, the mechanism of the explosion was as follows:

– regardless of the location of the point at which the ignition of the hydrogen–chlorine mixture started, two flame fronts began to propagate in the deflagration regime towards each end of the chloroduct, accelerating under the effect of the expansion of the H_2O and HCl gases produced by the reaction;

– assuming a hydrogen–chlorine mixture composition was slightly greater than the LDL, the acceleration was sufficient for each flame front to travel the pre-detonation distance, and a detonation has been established;

– the literature does not provide precise information on this distance or on how to evaluate it, especially since it depends on many parameters including at least the composition of the mixture and the diameter of the chloroduct (e.g. a modeling carried out for a stoichiometric mixture contained under 1.4 bar in a chloroduct of 200 mm diameter leads to a pre-detonation distance of 46 m);

– the special conditions necessary for the development of a pressure well above the pressure P_{det} = 30.7 bar lie in the fact that the geometry of the chloroduct was particularly well suited to the occurrence of a phenomenon similar to that of a "Double Discontinuity Detonation" as defined by Médard (1999):

- there would have been a pre-compression of the mixture, then the propagation of a detonation wave in that pre-compressed mixture and finally the reflection of that wave on the downstream closed end of the chloroduct, in the form of a shock wave that is more intense and faster than the incident wave and that catches up with the detonation wave propagating towards the opposite end of the chloroduct,

- it is then possible that the pressure developed locally has reached its rupture pressure P_r.

The TNT equivalent of the chloroduct burst can be estimated by applying the Brode relationship:

$$Eq_{TNT}=1/(4.69\ .\ 10^6)\ (P_r\text{-}P_0)/(\gamma\text{-}1)V \qquad [C18.5]$$

where:

– Eq_{TNT} is the TNT equivalent of the burst (in kg);

– P_r is the burst pressure of the chloroduct (in Pa) or 19,449,600 Pa;

– P_0 is the atmospheric pressure or 101,300 Pa;

– V is the total volume (in m^3) of the three broken sections 64 m long of the chloroduct or V = 2.4 m^3;

– γ is the ratio of the specific heat of the hydrogen–chlorine mixture, i.e. γ = 1.4;

– 4.69 . 10^6 is the value of the energy released by detonation of 1 kg TNT (in J/kg).

The calculation leads to Eq_{TNT} = 25 kg.

Strictly speaking, it is to be expected that the Brode relationship will overvalue the TNT equivalent of the explosion, since the bursting did not result from the action of a uniform pressure throughout the 64 m of the exploded chloroduct (as would be the case with a pneumatic bursting), but of the action of a pressure front reaching this value locally.

This is what emerges from a comparison between the value of Eq_{TNT} and the E_{eff} energy, which is deduced from the effects of the burst: the peak of the aerial overpressure it produced was estimated between 20 and 30 mbar at 80 m. However, according to the abacus TM5-1300 as shown in section 2.5.1.1.4, such an overpressure level corresponds to the detonation of only 4 kg of TNT.

Lessons learned from investigation of the explosion of a chloroduct

In order to identify the mechanism of the explosion, the chemical reactivity of the steel with chlorine and related compounds were examined:

– the formation of $FeCl_3$ on the inner wall of the chloroduct can be considered as a form of passivation of steel against its subsequent attack by chlorine, but this is only true if $FeCl_3$ does not give rise to the formation of the hexahydrate ($FeCl_3.6H_2O$): indeed, this hydrate is fused at a relatively low temperature (37°C) and its melting product is a corrosive liquid that quickly attacks the steel with hydrogen release;

– this means that it is essential to avoid the introduction of water into a chloroduct, including in the form of water vapor, and that periodic endoscopic checks must also be carried out to ensure that no moisture is deposited on the wall of a chloroduct;

– in any case, it is preferable to avoid heating a chloroduct or to do so only after ensuring the absence of a layer of hydrate on its wall;

– the monitoring of the surface aspect of the inner wall of a chloroduct is made all the more necessary as the presence of hydrate constitutes a potential permanent source of hydrogen, whatever the conditions of the formation of this hydrate (slow or fast speed and whether this presence is recent or old);

– finally, when the operation of a chloroduct is stopped, it must be avoided to keep it under chlorine before isolating it; it is preferable to purge it of the chlorine it contains, for example by nitrogen, before isolating it, or better to leave it under a low flow of nitrogen.

Case 19

Combustion of Steel in Oxygen

C19.1. Case presentation

A very violent combustion of steel in contact with highly pressurized oxygen, occurred in a facility delivering pressurized oxygen from cylinders.

C19.1.1. *Description of the facility*

The high-pressure part of the oxygen delivering facility consisted of the following elements, from upstream to downstream (see Figure C19.1).

Figure C19.1. *Scheme of the facility*

– a valve V1 was connected by a flexible hose, to the valve of a bundle of oxygen cylinders under high pressure (up to 200 bar);

– a pipe in stainless steel (diameter 34/26 mm), approximately 40 m long, which connected V1 to a valve V2;

– downstream V2, the pipe was connected, via a valve V3, to the following elements, attached to a metallic panel fixed to a concrete wall (all the pipes connecting these elements were in stainless steel):

- a four-way cylindrical fitting, directly connected to V3,
- a vent, via a valve V4,
- a Bourdon manometer.

– a pressure regulator equipped with a filter, connected to the low pressure circuit of the oxygen delivering facility.

Considering their constitution, V1, V2, V3 and V4 were ball valves, with a PTFE seat. V1 and V2 were in stainless steel, while V3 and V4 were in carbon steel. The sealing between the body of each valve and its upstream and downstream pipes was obtained by means of screwed tips, with metal–metal seals.

The pipes connecting V2 to V3, as well as the four-way fitting connected to V3, to V4, to the manometer and to the regulator, were in stainless steel. The fitting was welded to each of these four pipes.

The position, diameter and length of each pipe are given in Table C19.1.

Position of the pipe	**Diameter (mm)**	**Length (m)**
Connecting V2 to V3	15/21	1.5
Connecting V3 and the fitting	15/21	0.15
Connecting the fitting and the regulator	15/21	0.15
Connecting the fitting and V4	10/14	0.2
Downstream V4, coming out from a canopy, above the panel	10/14	2
Connecting the fitting to the manometer	4/6	0.3

Table C19.1. *Position and characteristics of the different pipes of the facility*

C19.1.2. *Circumstances of the accident*

The assembly of valves and pipes had been recovered from another site and, before its commissioning, the following operations had to be done:

– to install V1, V2 and the pipe connecting V2 to the elements of the panel;

– to close all the valves;

– to open V1 in order to pressurize the facility with oxygen under 150 bar, up to V2, and to test the sealing of the pipe between V1 and V2;

– to open V3 in order to check the sealing downstream V2, by pressurizing the pipe between V2 and V3.

During the latter operation, although V3 was closed, the manometer indicated an abnormal increase of the pressure downstream V3, which could only result from a default of its sealing.

The accident occurred just as an operator began to open V4, in order to vent the oxygen downstream V3.

C19.1.3. *Effects*

The witnesses of the accident were three operators who stood near the panel. They agreed to observe two very rapid phenomena, a weak explosion noise, followed immediately by a very bright and intense flame coming from the panel.

The accident has produced mainly thermal effects:

– the flame seriously burned the operator standing in front of the panel and slightly the two others standing behind him;

– the flame was hot enough to melt the steel of several elements of the panel, and then to initiate a combustion, so that a total mass of steel estimated to 1 kg was consumed.

The steel of the following elements has been partially or completely consumed:

– the lower half of V3 (body, seat, ball, seal);

– the pipes connecting V3 to the four-way fitting and this fitting to the regulator (the latter included copper elements that remained intact);

– some parts of the pipes connecting the manometer and V4 to the fitting, which were the closest to this fitting.

Other elements that were not in contact with oxygen were also marked by the flame:

– in several locations, the surface of the panel showed black burns or melted metal projections;

– some metal parts, such as metallic collars fixing the valves to the panel and the manometer cover showed traces of superficial fusion.

C19.1.4. *Mechanism of the accident*

C19.1.4.1. *Combustion of Fe in oxygen*

The combustion of Fe produces different oxides:

– FeO (ferrous oxide) for which the oxidation degree of Fe atom is (+2);

– Fe_2O_3 (ferric oxide) for which the oxidation degree of Fe atom is (+3);

– Fe_3O_4 (magnetic oxide).

In the following, the combustion of Fe in oxygen, is supposed to produce Fe_3O_4, according to the reaction [C19.1].

$$3\ Fe + 2\ O_2 = Fe_3O_4 \quad [C19.1]$$

The enthalpy of formation of Fe_3O_4 equals -1.121 MJ/mole of oxide.

From this value and from the thermal molar capacity of Fe_3O_4, i.e. 143 J/mole.K, the calculation leads to the adiabatic flame temperature T_f corresponding to the stoichiometric combustion of Fe in oxygen, i.e. more than 8,000 K.

C19.1.4.2. *Application to oxygen thermal lance*

The principle of the oxygen thermal lance is based on the very high value of T_f:

– a lance is a steel tube opened at both ends, in which oxygen flows and where the combustion of steel is initiated at the outlet end;

– once combustion has started, it continues with the consumption of steel from the tube;

– the reaction zone is raised to a very high temperature, as well as its close environment;

– although the combustion conditions are not adiabatic, the temperature of the reaction zone is greater than 3,700 K, which is higher than the melting temperature of many materials.

Hence, the lance is used as a tool for cutting or drilling hard to fuse materials (steel, concrete, slag, etc.).

C19.1.4.3. *Application to the accident*

During the accident, the steel of the elements downstream V3, which were in contact with oxygen under high pressure, was involved in a combustion similar to that in an oxygen lance.

Then it is necessary to determine:

– the sequence of events that allowed these elements to be in the presence of oxygen under high pressure;

– the mechanism through which the combustion of steel was initiated and propagated.

C19.1.4.3.1. Presence of oxygen under high pressure downstream V2

When the operator opened V2 to pressurize the pipe between V2 and V3, the indication of the manometer meant that the pipe downstream V3 had also been pressurized, because V3 was leaking.

C19.1.4.3.2. Mechanism for initiating and propagating the combustion of steel

One of the circumstances of the accident results from the following events:

– before the pipe between V1 and V2 had been pressurized by oxygen, its two end fittings, connected to V1 and to V2, had been replaced by more resistant ones;

– for this replacement, the old fittings were first laid out from the pipe by sawing, then the new ones were welded;

– the sawing and welding operations had been carried out on the pipe maintained in place, so that the steel dust (swarf) produced during the sawing dropped vertically within the pipe end, upon the ball of V2, and remained there until it fell under, when V2 was opened.

The combustion of steel has probably resulted from the following mechanism:

– before the operator began to open V4, the V3 leak allowed the pipe downstream V2 to be under oxygen pressure and the steel dust was present at rest under the ball of V2;

– when V4 was opened, the grains of this dust were blown off and transported downstream up to the four-way fitting and to V4;

– they had just been formed and were very oxidizable, so that their combustion has been initiated;

– this combustion propagated to a massive element of the facility, probably the four-way fitting, then this element broke through and the total amount of oxygen contained in the facility was released, feeding the combustion of other massive elements.

The internal volume of the facility which contained oxygen under 150 bar can be estimated to 21 L. This volume represents a mass of oxygen approximately equal to 4 kg, which, according to the reaction above, could react with nearly 11 kg of steel, in stoichiometric conditions.

This figure is consistent with the total mass of steel consumed in the combustion of the various elements of the facility, which can be estimated to about 1 kg.

Lessons learned from investigation of combustion in oxygen

– Oxygen under high pressure is a powerful oxidizer able to initiate, spontaneously and even at ambient temperature, the violent combustion of various materials. These include not only easily oxidizable products such as hydrocarbon ones (elastomeric seals, greases, etc.) but also certain metals, including steel, when it is finely divided as dust.

– When steel starts to burn as a massive material, the combustion can continue as long as the reaction zone is supplied with oxygen.

– Companies that distribute oxygen must specify the different precautions to be taken, when oxygen under high pressure is implemented.

Case 20

Explosion in an Aluminum Foundry

C20.1. Case presentation

The mechanism of an explosion that completely destroyed the casting facilities of an aluminum foundry had to be determined.

C20.1.1. *Description of the facility*

The facility consisted essentially of a set of several casting machines, each one being able to produce simultaneously, in a semi-continuous casting process, six aluminum plates, of approximately 0.4 m x 1.1 m cross-section and 2.9 m high, according to the following procedure:

– aluminum is kept liquid above 700°C in a retention furnace;

– by tilting this furnace, the metal is poured into a duct that channels it to vertical nozzles;

– each nozzle feeds liquid aluminum into the top of a plate whose edges have already solidified and play the role of a container;

– water run-off on the vertical sides of the plates ensures the cooling and progressive solidification of the liquid aluminum;

– the plates are fixed to a vertically guided, two-rail descender which, during the casting process, gradually descends into a 10-meter deep well containing water up to half-height;

– the speed of the descender, about 50 mm/min, and the flow rate of the liquid aluminum are adjusted so that the level of the liquid surface remains constant and corresponds practically to the level of the ground;

– at the end of the casting process, the pouring of aluminum at the top of the plates is stopped, the descent of the casting machine is also stopped and shortly thereafter, the water run-off on the plates is interrupted.

C20.1.2. *Circumstances of the explosion*

At the time of the explosion, a cast of six aluminum plates was in progress and, from the height of their solidified part, the end of the casting process was very close: the pouring of the liquid aluminum had already been stopped and a total mass of between 600 and 1,200 kg, not yet solidified, remained at the top of the plates, while water run-off was still in progress.

All the witnesses agree that the explosion occurred during a thunderstorm, and more precisely when lightning struck the foundry building, a fraction of a second before the explosion.

C20.1.3. *Explosion effects*

The explosion killed four operators, three of them were right next to the well. It has also devastated the facilities by producing several types of damage.

C20.1.3.1. *Damage to the casting well*

Its four vertical walls have been extensively deformed since they have all been pushed from several centimeters to several decimeters depending on the height, after the iron rods of the reinforced concrete have broken and the pairs of walls of each dihedron have separated.

At 5 m below ground level, i.e. slightly above the assumed level of water in the well, the deformations are the most important, resulting in the presence of a concavity of the walls.

The explosion was violent enough to be detected by seismographs present at distances of several hundreds of kilometers.

C20.1.3.2. *Damage to facilities in the casting well*

The cable supporting the descender and the six plates being casted were broken in several places, resulting in the assembly falling into the well.

Two welded U-irons, which constituted the upper portion of the descender, were crushed. Similarly, pipes of several millimeters thick arranged vertically in the well were flattened at the top of the well and at the water level.

All this damage shows that the pressure in the casting well reached at least 1 bar, which corresponds to the application, on its walls, of a force reaching at least 100 daN/m^2.

C20.1.3.3. *Damage to the plates being casted*

The debris from the plates fell to the bottom of the well:

– at the time of the explosion, the upper cavity of each one was filled with liquid aluminum and, for four of the six plates, the edges of the cavity were fractured into several pieces;

– three of the plates were fractured into two pieces in a straight section at the top quarter.

C20.1.3.4. *Damage to other casting installations*

All the elements of the casting machine (ingot molds, casting ducts, water boxes and their supports) have been broken up into many very deformed pieces. The same is true of the decking parts, which were 60 cm below ground level. These various pieces of debris were projected at varying distances (up to 700 m), in a preferred direction, with a greater concentration of debris being found at an angle of 90°, centered on the northeast of the well.

Figure C20.1. *View of damage to all the foundry facilities*

All other installations in the vicinity of the well were destroyed (driver's cab, casting duct, etc.). The retention furnace moved away at more than 1 m from the casting well (see Figure C20.1).

C20.1.3.5. *Damage to foundry buildings*

In its northern part, the foundry building was completely destroyed and the adjoining buildings were severely damaged (see Figure C20.2).

Figure C20.2. *Aerial view from the north of the foundry building*

The north gable of the foundry was approximately 30 m from the casting well. The cinder block walls of this gable and those of the two adjoining buildings to the east and west were blown. Different rooms were built on two levels and leaned against the north gable of the foundry: their upper level was completely blown and the lower level was severely damaged.

The foundry building was covered with a roof consisting of a metal frame covered by a light cladding: within a radius of 30 m centered on the axis of the well, the roof was completely blown.

At a greater distance, the trusses of the frame were deformed but only the cladding was blown away. A greater amount of the roof plates was found more than 1 km northeast of the well, which may be explained by the fact that it was subjected to the influence of a southwest wind at the time of the explosion.

C20.1.3.6. *Damage to other plant buildings*

The cinder block walls of the laboratory, located between 60 and 80 m north of the well, were cracked in several places.

All the doors, windows, ceilings and partitions of the rooms of this building have been blown.

For more distant buildings, damage is limited to torn window frames or broken roof windows or canopies.

As with casting debris, a greater concentration of broken glass and broken window frames is located at a 90° angle centered on the north of the casting well. This is why we admit that, in all directions, it is up to maximum distances R of the center of the explosion between 400 and 600 m that windows were broken. Similarly, it is up to maximum distances between 200 and 250 m that window frames were torn off.

C20.1.4. *Determination of the mechanism of explosion*

C20.1.4.1. *E_{eff} estimation*

In accordance with the method described in Chapter 2 (see section 2.3), the determination of the mechanism of explosion begins with an estimation of E_{eff}, using the TNT equivalent model (this model is relevant since the energy of the explosion was released at the location of the casting well, i.e. very locally compared to the distances where damage was produced):

– the value of Eq_{TNT}, expressed in kg, is derived from the values of the reduced distance λ corresponding to each type of damage produced at the distance R from the casting well, which is the center of the explosion, in application of the relationship (see section 2.5.1.1.3 of Chapter 2):

$$Eq_{TNT} = (R/\lambda)^3$$

– a window frame is torn off by an aerial wave whose overpressure peak equals $\Delta P = 50$ mbar; according to the abacus TM5-1300, this pressure value corresponds to $\lambda = 25\ m/(kg)^{1/3}$ and for R between 150 and 200 m, the above relationship leads to Eq_{TNT} between 200 and 500 kg; similarly, a glass is broken for $\Delta P = 20$ mbar, which corresponds to $\lambda = 70$ m/(kg)$^{1/3}$ and for R between 450 and 550 m, the relationship leads to Eq_{TNT} between 250 and 500 kg;

– ultimately, damage analysis leads to the following values:

- 200 kg < Eq_{TNT} < 500 kg,

- 900 MJ < E_{eff} = 4.69 Eq_{TNT} < 2,400 MJ.

Such values of Eq_{TNT} and E_{eff} characterize a very violent explosion. In order to identify the explosive system responsible for it, both the explosive characteristics and the quantity of the products involved must be taken into account:

– the foundry was equipped with a network of pipes for supplying the furnaces with natural gas. The hypothesis of an ATEX explosion (explosive system 1), which was formed as a result of a gas leak from this network, must be considered, but it can be eliminated because it does not correspond to the intensity or location of the damage suffered by the casting well; moreover, the explosion should have occurred in the regime of rapid deflagration, but the geometry of the foundry building did not promote it;

– the explosion did not involve the bursting of a hollow body (explosive system 3) and this hypothesis can therefore also be ruled out;

– moreover, the water could have vaporized abruptly in contact with the molten aluminum (explosive system 4), but this contact led to a chemical reaction rather than a simple vaporization which would produce an energy that is too weak;

– finally, it is the hypothesis of a chemical reaction involving aluminum (explosive system 2) that must be preferred.

C20.1.4.2. *Hypothesis of an explosion involving aluminum*

The nature of the reagents present suggests that the explosion resulted simultaneously from the combustion of aluminum in the air and its reaction with water. In the case of the latter reagent, it was present in the vicinity of the liquid aluminum, both in the form of water flowing down the plates and flowing also through the pipes feeding the casting line.

C20.1.4.2.1. Dispersion of liquid aluminum

To allow the combustion of aluminum and its reaction with water, it was first necessary that liquid aluminum be dispersed in the air and it is lightning that is at the origin of this dispersion:

– although the building was protected from lightning according to the current standard, the lightning current flowed to the ground using the liquid aluminum as an electrical conductor;

– the liquid aluminum was instantaneously heated to a very high temperature so that it was partly aerosolized with fine droplets dispersed into air and probably also partly vaporized;

– at the same time, it is likely that the dispersion of aluminum had the characteristics of a real explosion, with the production of mechanical effects which caused the breakdown of certain elements of the casting installations, and caused the formation of an aerosol of fine droplets of water suspended in the air.

C20.1.4.2.2. Combustion of aluminum

The complete combustion of aluminum, very exothermic, is a reaction between 2 moles of aluminum and 1.5 moles of oxygen, which, in the proportions of air, is associated with 5.66 moles of nitrogen; this first reaction [C20.1] is therefore written as (its enthalpy ΔH is worth - 45.2 MJ/kg Al):

$$2\ Al + 1.5\ O_2 + (5.66\ N_2) = Al_2O_3 + (5.66\ N_2) \qquad [C20.1]$$

C20.1.4.2.3. Reaction of aluminum with water

The reaction of aluminum [C20.2] with water needs 2 moles of aluminum for 3 moles of water and is also exothermic, although three times less than combustion (its enthalpy ΔH is worth - 15.2 MJ/kg Al):

$$2\ Al + 3\ H_2O = Al_2O_3 + 3\ H_2 \qquad [C20.2]$$

C20.1.4.2.4. Reaction of hydrogen with oxygen in air

It is also necessary to consider that the hydrogen produced in the reaction of aluminum with water can react with the oxygen in air, according to the reaction [C20.3] (its enthalpy ΔH is worth - 143 MJ/kg H_2):

$$H_2 + 0.5\ O_2 + (1.89\ N_2) = H_2O + (1.89\ N_2) \qquad [C20.3]$$

This third reaction is also exothermic, but its influence is certainly negligible here, since H_2 is one of the reagents and it is also one of the products of the second reaction. The third reaction would therefore have a much lower degree of progress than those of the first and second reactions, which involve reagents already present in the medium.

C20.1.4.2.5. Explosion yield

Whether the explosion was the result of combustion of the aluminum aerosol in air or a reaction of the aluminum aerosol with the water aerosol, it can be said that the overall yield of the explosion was significantly less than 1, rather close to 0.2, taking into account the poor quality of the mixture of reagents and their physical state, aerosol for aluminum and water and gaseous for oxygen.

This resulted in the reaction releasing an E_{lib} energy of between 4,500 and 11,500 MJ, and the necessary quantities of reagents were as follows:

– assuming that the explosion involved the combustion of aluminum alone:

- between 100 and 250 kg of liquid aluminum,

- between 300 and 750 m^3 of air;

– assuming that the explosion involved the only reaction of aluminum with water:

- between 300 and 750 kg of liquid aluminum,

- between 300 and 750 kg of water.

C20.1.4.2.6. Hypothesis of a double reaction

Based on the circumstances of the explosion, the minimum aluminum and water requirements of 750 kg for each of these reagents were met. With regard to air, based on the high value of the minimum volume, between 300 and 750 m^3 (which corresponds, for example, to a cube with an edge between 6.7 and 9.1 m), the conditions for the rapid formation of a sufficiently homogeneous mixture were difficult to satisfy, and the hypothesis of an explosion involving the combustion of aluminum alone is therefore unlikely.

For an identical release of energy, the reaction of aluminum and water aerosols requires three times more aluminum than the combustion reaction, but the rapid formation of a sufficiently homogeneous mixture is perhaps easier, since the same cause generated both aerosols.

Finally, we accept that the explosion involved both the combustion of the aluminum aerosol and its reaction with the water aerosol, but it is not possible to specify the respective proportions of each of them.

C20.1.4.3. *Synthesis of the mechanism*

The explosion resulted from the sequence of the following events:

– the lightning fell on the foundry building;

– the lightning current has flowed to the ground, using the electrical conductor consisting of the liquid aluminum (between 600 and 1,200 kg) present at the ground level, at the top of the casting well and also at the top of the six plates being casted;

– the liquid aluminum was instantaneously and violently dispersed into air as an aerosol and was also likely partially vaporized;

– this violent dispersion caused mechanical effects that destroyed the casting machine and allowed the cooling water of the plates to be dispersed in the form of a water aerosol;

– the aluminum aerosol was burned in the air and reacted with the water aerosol;

– these two simultaneous reactions released a very high energy and generated an aerial overpressure wave that caused damage up to significant distances;

– from the analysis of the damage, the TNT equivalent of the explosion has been estimated between 200 and 500 kg. It is likely that the reactive mixture (air and both aerosols) was contained in a large part of the volume of the well because it is there that the explosion reached the highest level of pressure, i.e. 1 bar, and this level even suggests that, in the well, the regime of explosion was the detonation.

Lessons from investigation of an explosion in an aluminum foundry

The explosion involved the combustion of liquid aluminum in the form of an aerosol heated to very high temperatures.

It also involved the reaction of the same aluminum aerosol with a water aerosol, the formation of which resulted from a dispersion of liquid aluminum and water as a result of the mechanical effects of the lightning current.

The mechanism of the explosion was determined by the method described in Chapter 2 of this book, the first step being an estimation of the energy released to produce the observed damage.

Case 21

Explosion in a Laboratory Nitration Test

C21.1. Case presentation

In a laboratory, a 3-liter cylindrical test vessel exploded, some minutes after an operator introduced an aqueous solution of nitric acid into the vessel already containing an aqueous solution of glyoxal.

The operator was fatally injured by this violent burst, the mechanism of which had to be determined.

C21.1.1. *Nature of the explosive system*

The circumstances of the explosion indicate that the explosive system was formed when the two solutions were mixed in the vessel:

– an exothermic reaction was initiated between the concentrated nitric acid solution (mass content = 68%), which is a powerful oxidizer, and the aqueous solution (mass content = 40%) of glyoxal ($C_2H_2O_2$), which is a very oxidizable hydrocarbon dialdehyde, for which the degree of oxidation of carbon is 0;

– it was an oxidation–reduction reaction, giving rise to a runaway which can be written as the following equation [C21.1]:

$$5\ C_2H_2O_2 + 6\ HNO_3 = 10\ CO_2 + 3\ N_2 + 8\ H_2O \qquad [C21.1]$$

Based on the thermochemical properties of the reagents, the characteristics of this reaction are as follows:

– the enthalpy ΔH is worth - 13.7 kJ/mole of glyoxal;

– the stoichiometric composition corresponds to a molar proportion nitric acid/glyoxal = 55/45;

– it is a priori for such a composition, or at least very close to stoichiometry, that the heat of reaction is the highest and therefore the thermal runaway gives rise to the most violent effects (in fact, the burst resulted from an operator mistake in the proportion of the two reagents and, for nominal proportions, no runaway would have occurred);

– in addition, for the temperature of the reaction, all the products are gases and, for 1 mole of glyoxal, the total number of moles of these gases has a high value close to 4.

C21.1.2. *Experimental validation of the conditions of the runaway reaction*

A series of tests was carried out to demonstrate that the total mass of the solutions introduced into the vessel can develop a pressure sufficient to cause its burst. These tests, which covered various volumes of aqueous solutions of nitric acid and glyoxal, and various compositions close to the stoichiometry, consisted of:

– remote controlling the introduction of a known volume of nitric acid solution into the vessel already containing a known volume of glyoxal solution, having a concentration equal to or slightly less than 40%;

– recording the variations of the temperature of the liquid phase T_L, of the temperature of the gaseous phase T_G and of the internal pressure P.

For this series of tests, the following parameters were varied:

– the diameter d of a diaphragm fitted with one of the three dome tubings (the two others and the drain hole having been closed) varied between 2 and 10 mm;

– the total mass of the mixture of the two components varied between 0.2 and 1.2 kg;

– the possible presence of formaldehyde CH_2O in the aqueous solution of glyoxal, at 5% by weight (in the circumstances of the accident, formaldehyde was present in small amounts in the glyoxal solution and some of the tests were therefore carried out in the presence of this product).

C21.1.3. *Results*

The analysis of the results of the dozen tests carried out shows that:

– a chemical reaction occurs between the two reagents introduced into the vessel and results in a more or less intense and more or less rapid increase in the T_L and T_G temperatures, as well as the pressure P;

– this reaction starts after an induction delay, ranging between a few dozen seconds and more than 1 minute; this delay, as well as the maximum values of T_L, T_G and P were higher in the presence of formaldehyde;

– for each d value, Table C21.1 gives the maximum values of T_L, T_G and P, which were obtained with a glyoxal solution containing formaldehyde and for a total mass of reagents = 1.2 kg;

d (mm)	**Induction time (s)**	**Measured maximum values**		
		T_L (°C)	T_G (°C)	P (bar)
10	40	159	144	3
8.5	40	224	219	23
5	45	190	NA	125
2	65	>200	NA	NA

Table C21.1. *Maximum values of T_L, T_G and P, measured for different d values (NA = not available)*

– Figure C21.1 shows the variation of T_L, T_G and P obtained for d = 8.5 mm:

- the variations of T_L and T_G show quite similar rates of temperature rise, averaging about 50°C/s between 20 and 200°C,

- the variation of P shows a maximum pressure close to 23 bar which was not sufficient to reach the plastic deformation of the vessel, and a maximum rate of pressure rise of several dozen bar/s;

– for d = 5 mm, the pressure reached 125 bar, and although the container did not explode, its rounded aspect shows that the elastic limit of the steel was exceeded;

– during the test corresponding to d = 2 mm, the vessel exploded:

- the maximum values of T_L and T_G were too high and reached too quickly to be able to estimate them,

- the value of the rupture pressure P_r of the vessel could not be measured.

Based on the mechanical characteristics of the vessel, it is possible to estimate P_r:

– the vessel had a diameter D = 200 mm, and was made of a sheet of stainless steel with a thickness e = 3 mm; it was equipped with three welded tubings, approximately 10 mm in diameter, spread over its dome, as well as a drain hole;

– according to Lamé's relationship applied to a cylinder with the same diameter and thickness as those of the vessel, it is possible to calculate its rupture pressure P_r:

$$P_r = 2e.\sigma_r/D \quad [C21.2]$$

– the rupture stress σ_r of the steel is not known precisely but it can be estimated at 500 MPa, so that the calculated value of P_r is worth 150 bar;

– in fact, it is likely that the value of P_r was less than 150 bar, since the vessel was made of welded sheets (ferrule, bottom and dome) and these welds are fragile zones;

– during the test with d = 5 mm, the vessel resisted to an internal pressure of P = 125 bar;

– then, the rupture pressure P_r can be estimated to approximately 130 bar which is consistent with the previous values.

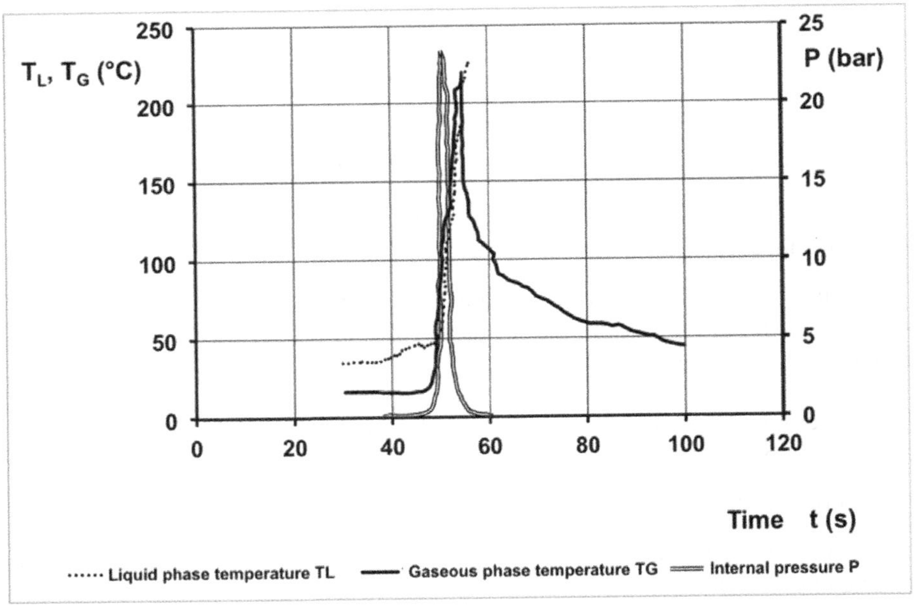

Figure C21.1. *Variation of T_L, T_G and P obtained for d = 8.5 mm*

The burst produced three pieces (see Figure C21.2).

Figure C21.2. *View of the three pieces of the burst vessel*

Their aspect, as well as their limited number, indicates that the explosion did not result from a detonation of the explosive system but from a deflagration and that the burst of the vessel resulted from an increase of the pressure of the gases released by the reaction, until it reached its rupture pressure P_r.

C21.1.4. *Conclusion of the tests*

In the end, the mean values of the rates of temperature and of pressure rise, the existence of a 40-second induction delay during which there was no significant increase in the T_L, T_G and P parameters and the general aspect of the curves of Figure C21.1 reflect a phenomenon of runaway.

Lessons learned from investigation of a burst vessel

– The experience shows that an oxidation–reduction reaction can result in a very severe runaway, provided that the enthalpy of the reaction is sufficiently high: it is such a runaway that caused the burst of the vessel;

– the proportion of each reagent is a parameter that significantly affects the reaction speed: the most rapid reaction corresponds to the stoichiometric proportion;

– an extra-reagent (formaldehyde in this case) may have a significant influence on the reaction characteristics (induction delay, reaction speed, maximum temperature and pressure of the products), even if the concentration of this extra-reagent is small;

– in the burst test, the vessel was equipped with a diaphragm, but its diameter was too small to prevent the burst;

– this situation may occur for processes operated with reactors of several m^3 under industrial conditions;

– it is essential to protect a chemical reactor from the risk of runaway. Generally, this protection requires a pressure release device, which must allow a sufficient reduction of the pressure inside the reactor, in order to keep it under its rupture pressure. This device may be a diaphragm, a rupture disk or a pressure release valve. Its diameter should be dimensioned in a very rigorous way, which implies knowing the chemical reactivity of the reaction mixture and using suitable models and guidelines.

Case 22

Explosion in a Chemical Reactor

C22.1. Case presentation

An explosion occurred in a phthalic anhydride synthesis unit and its mechanism had to be investigated.

C22.1.1. *Description of the chemical synthesis process*

Anhydride phthalic ($C_8H_4O_3$) is produced by a mild oxidation of o-xylene, according to this slightly exothermic reaction [C22.1]:

$$C_8H_{10} + 3\ O_2 = C_8H_4O_3 + 3\ H_2O \qquad [C22.1]$$

The unit consisted of two identical cylindrical tubular-beam reactors, operating in parallel as heat exchangers. In this type of reactor, the reaction mixture is a gaseous mixture of air and o-xylene vapor, which is introduced into the upper part of the reactor, under pressure close to atmospheric pressure, and which flows within the 16,000 vertical tubes of the beam (each tube is 2.4 m high and has a 20 mm internal diameter and is welded by its ends to two horizontal plates).

The mild oxidation of o-xylene occurs when a catalyst is deposited in a fixed bed on a silicon carbide support contained in the tubes. The reactor mixture is then introduced into a cooler fitted to a battery of condensers in parallel, in which phthalic anhydride is separated from other waste products of the reaction (water vapor and a residual of o-xylene).

The heat produced by the reaction is transferred to a heat transfer liquid that circulates between the tubes and consists of a mixture of melted salts (nitrites + alkaline nitrates). At the starting of the unit, this liquid is heated by electrical resistances and the heat is transferred to the reaction mixture, so as to initiate the reaction. Then, the temperature of the reactor is maintained between 400 and 450°C, i.e. the interval within which the yield is maximum.

The heat transfer liquid circulates in a closed circuit that essentially consists of:

– the two reactors;

– a circulation pump;

– a heat exchanger which, when the unit is started, ensures the heating of the heat transfer liquid and which, when the reaction has been initiated, ensures the evacuation of the heat transmitted by the reactors to the heat transfer liquid.

C22.1.2. *Circumstances of the explosion*

The explosion occurred during the starting of the unit. This was carried out under unusual operating conditions since only one reactor was started. The internal temperature sensor of the reactor quickly drifted and indicated an abnormally high value. Then, the internal pressure of the reactor increased until an alarm in the control room was triggered, projections of the melted salt were released from the gage wells and white fumes containing red nitrous fumes were emitted at various locations in the unit. Then, four minutes after the alarm was triggered, the reactor exploded.

C22.1.3. *Effects of the explosion*

The top plate of the tube beam was broken along the diameter, and the cylindrical wall of the reactor was torn along its wall. The reactor opened widely, showing the inside zone close to its axis: in the zone below the top plate, the tubes disappeared by consumption of steel and a solid brown product was formed. On the contrary, the tubes located on the periphery of the reactor are intact or have been stretched and some have traces of steel melting. All of the observed effects suggest that the burst involved the core of the reactor.

The explosion also caused damage to the reactor's immediate environment: the roof of the building that housed it, as well as its northern gable, was destroyed or severely damaged.

The very hot melted salt projections made of a very oxidizing liquid caused the ignition of the fatty fuel products present on the soil at the different levels of the building, causing a widespread fire.

The air pressure wave resulting from the explosion caused damage at a greater distance (broken glass or roof slabs, broken cladding, etc.).

Finally, a lot of debris was thrown, some of which was about 300 m away.

From the mapping of the observed damage, the TM5-1300 abacus has been used to estimate the TNT equivalent of the explosion between 150 and 200 kg.

C22.1.4. *Determination of the explosion mechanism*

To determine the explosion mechanism, two hypotheses must be successively evaluated, which both take into account the thermochemical properties of the products in presence.

C22.1.4.1. *Hypothesis of a combustion of o-xylene*

O-xylene is a flammable liquid with a flash point FP = 32°C; the values of its LEL and UEL are, respectively, 1% and 7% volume (INRS ED 911 2004). The o-xylene content of a stoichiometric air–o-xylene mixture is 2% volume; the combustion reaction [C22.2] of this mixture is written as follows and has an enthalpy close to -42.9 MJ/kg:

$$C_8H_{10} + 10.5\ O_2\ (+\ 39.6\ N_2) = 8\ CO_2 + 5\ H_2O\ (+\ 39.6\ N_2) \qquad [C22.2]$$

Under the normal operating conditions of the reactor, the air–o-xylene mixture used has a composition that belongs to the flammability range and it could be assumed that the explosion involved the combustion of o-xylene in one or more tubes of the reactor.

Based on the value of the o-xylene combustion enthalpy, the energy released by the combustion of a stoichiometric mixture contained in a reactor tube is worth 3 J, assuming that the tube is fully filled with the mixture (this is a major hypothesis, since it supposes that the volume occupied by the medium catalyst is negligible).

Assuming that the combustion occurred in all 16,000 tubes of the reactor, it would have released a total energy $E_{lib} = 48$ MJ, which represents an equivalent TNT $Eq_{TNT} = 10$ kg. This value is 15–20 times lower than the TNT equivalent deduced from the explosion damage. Moreover, the combustion of o-xylene could only develop in one tube a maximum pressure not exceeding 10 bar and thus quite insufficient to cause the tube to burst, since its bursting pressure can be estimated at several hundred bars. Finally, the hypothesis of a combustion of o-xylene can be ruled out.

C22.1.4.2. *Hypothesis of a reaction between nitrate and steel*

Molten salt contains alkaline nitrates that have recognized oxidizing properties, but only at temperatures significantly higher than the normal operating temperatures of the reactor.

It must therefore be supposed that, as a result of an abnormal temperature drift, nitrate has proved to be in a condition favorable to the appearance of an oxidation–reduction reaction between the nitrate as an oxidizer and a reducing product which remains to be identified.

The hypothesis has been put forward that this product is the Fe atoms contained in the steel (grade A179) of the reactor tubes.

The simplified writing of the oxidation–reduction reaction [C22.3] between sodium nitrate and Fe could be as follows:

$$15\ Fe + 8\ NaNO_3 = 5\ Fe_3O_4 + 4\ N_2 + 4\ Na_2O \qquad [C22.3]$$

For 1 kg of Fe, the enthalpy of this reaction is –4.6 MJ and is thus nearly equal to the enthalpy of decomposition of TNT, which characterizes a rather highly exothermic reaction that may be involved in a thermal runaway.

If we accept that the mass of the steel of the tubes that was consumed in the explosion is 7 T, the calculation leads to the energy likely to have been released by the above reaction i.e. $E_{lib} = 30{,}000$ MJ, which corresponds to an equivalent TNT $Eq_{TNT} = 7$ T.

Since the TNT equivalent of the explosion has been estimated to be between 150 and 200 kg based on the damage it caused, the examination of this hypothesis shows that:

– the explosion may have resulted from a runaway reaction between the steel of the reactor tubes and the nitrate salt;

– it had a low yield of only 2–3%.

Note that this low yield value is typical for this type of explosion where each reagent is in a different state and the reaction occurs on the surface of a solid reagent, in contact with a liquid reagent.

C22.1.4.2.1. Bibliographic data

According to the literature, the reaction between alkaline nitrates and Fe becomes violent if the temperature is significantly above the normal salt temperature in the reactor. In an accidental explosion in 1959 during a similar synthetic reaction, the reaction of the molten salt on the reactor tubes had already been questioned (Allen 1980).

C22.1.4.2.2. Experimental testing of a reaction between nitrate and steel

A series of tests has been carried out to determine the temperature threshold at which molten salt begins to react with steel and also the temperature threshold at which the reaction results in a thermal runaway. The results were as follows:

– steel only undergoes surface oxidation as long as the salt temperature remains below 820°C;

– from 840°C, the reaction of the salt on the steel starts and is very exothermic since it heats the salt to more than 1,100°C, resulting in the melting of the steel;

– for this temperature, the reaction is fast enough to maintain itself, so that it is no longer necessary to heat the system.

C22.1.5. *Description of the explosion process*

The following elements were used to establish the explosion process:

– the explosion resulted from a runaway of the highly exothermic oxidation reaction between the steel of the reactor tubes and the nitrate contained in the molten salt;

– to initiate this reaction, the salt in the reactor had to be heated, at least locally, to more than 800°C;

– to explain a temperature drift from its nominal value of 400–450°C to more than 800°C, it must be admitted that the heat produced by the synthesis reaction of phthalic anhydride during the starting was not normally evacuated due to a failure in the circulation of salt;

– this failure may have had several causes but one of the possible ones is very likely: it is established that, during the maintenance operation that had taken place prior to the restarting of the production unit, the terminals of the three-phase electric motor that were driving the molten salt circulator had been reversed, which had also reversed the direction of rotation of the motor. This reversal of the flow is sufficient to explain a change in the thermal exchange conditions between the salt and the tubes in the reactor and finally an insufficient cooling of the tubes.

The explosion of the reactor was due to the fact that the oxidation reaction of the steel produced nitrogen and nitrogen dioxide NO_2, whose red smoke was observed by the witnesses. As a result of the temperature increase, the pressure in the reactor increased up to its rupture.

Lessons learned from the investigation

The mechanism of the explosion was established by comparing the TNT equivalent of the explosion based on the intensity of the damage and the energy likely to have been released in two possible chemical reactions, based on the thermochemical properties of each reagent and the quantities available.

The explosion was made possible by the failure of the salt circulation system. Such a device should be rigorously maintained and continuously monitored for its effectiveness by applying the following measures:

– it is essential to continuously monitor the temperature of the salt in the reactor, placing a sufficient number of probes in it, in particular on its axis, and to provide for preventive measures to stop the reactor from feeding reagents, to drain it and/or to cool its contents;

– before any restarting of the installation after maintenance, a measure should require the control of the direction of rotation of all three-phase electric motors whose power supply has been deposited during the maintenance.

Case 23

Explosion and Fire Resulting from an Oxidation by $KMnO_4$

C23.1. Case presentation

In the sugar industry, beet pulp is sometimes treated as a raw material: its fermentation transforms the sugar still present into ethanol, which is then extracted by distillation.

This alcohol contains sulfur-smelling products (mercaptans) and, to deodorize it, these products are at least partially removed by keeping the alcohol in contact with a powerful oxidizer for a sufficient time, in this case potassium permanganate ($KMnO_4$), in the state of solid crystals.

C23.1.1. *Circumstances of the explosion*

In a storage area of a sugar factory, comprising several tanks situated in a retention basin, explosions occurred successively in three tanks, in the following circumstances:

– several tanks of the storage area contained variable volumes of alcohol;

– to prepare a future sugar campaign, a tank dedicated to alcohol deodorization had been emptied of the alcohol it contained;

– on the bottom of this tank, there were solid crystals of $KMnO_4$, corresponding to the "old" product which had been present in the tank, over the previous year;

For a color version of all figures, in this case study, see www.iste.co.uk/chaineaux/mechanisms.zip.

– this “old” product had to be removed and replaced by a “new” one.

Then, the operators successively carried out the following actions:

– they put “new” crystals on the bottom of the tank;

– they introduced a small amount of alcohol into the tank;

– at this time, they left the site, planning to complete the filling the following day;

– about 30 minutes later, an explosion occurred in the tank, causing its destruction.

C23.1.2. *Effects of the explosion*

The effects of the explosion were as follows (see Figure C23.1):

– the roof of the tank was torn from the ferrule and projected around 100 meters away;

– its vertical welding was also torn apart, leading to the complete destruction of the tank;

– the alcohol contained in the tank spread into the retention basin and was ignited, causing a fire;

– before the fire department extinguished the fire, two other tanks, close to the first one and also containing alcohol, exploded too, the welding of their roof was also torn, but only on a part of the circumference.

Figure C23.1. *View of the destroyed tanks*

C23.1.2.1. *Mechanism of the explosion*

C23.1.2.1.1. Presence of ATEX in the tanks

Ethanol is a flammable liquid with a flash point FP = 12°C and an LEL = 6.3% volume. It can be supposed that the temperature of the atmosphere in each tank was close to the ambient temperature, and in any case above 12°C. Then, this atmosphere was a mixture of air and ethanol vapor, with a vapor concentration slightly above the LEL. Moreover, it can be supposed that, since this atmosphere was contained in a closed space, a liquid–vapor equilibrium had been established. Then, it is the ATEX present in the whole volume of each tank, which exploded.

C23.1.2.1.2. Identification of the ignition source of the ATEX in each tank

Ignition source of the ATEX present in the first tank

One laboratory test was carried out, in order to experimentally identify the ignition source of the ATEX in the first tank:

– a few grams of crystals of potassium permanganate $KMnO_4$ are placed in a test vessel and some ethanol is poured into it, in such a quantity that the liquid does not completely submerge the crystals, but leaves an emergent part;

– the ethanol ignition is spontaneously achieved within approximately 30 minutes.

This ignition results from the successive following two phenomena:

– the potassium permanganate oxidizes ethanol, strongly increasing its temperature;

– this heating is sufficient to initiate the self-ignition of the ATEX present as a mixture of air and ethanol vapor, above the reaction zone between the liquid ethanol and the emerged solid potassium permanganate;

– this mechanism can be extrapolated to the tank situation and explains how it caused the ignition of the ATEX in the first tank.

This tank was made of a ferrule welded to the roof and the mechanical resistance of such an assembly is relatively low, at most a few dozen millibars, which can be considered as its rupture pressure. The flame spread throughout the homogeneous atmosphere present in the sky, which was an ATEX since the concentration of the alcohol vapor was just above LEL. Then, it is possible to say that the explosion of this ATEX developed an overpressure sufficient to reach the rupture pressure of the tank and the projection of its roof. It must be noted that, if the tank had had a much higher rupture pressure, the explosion overpressure could have reached several bars.

C23.1.2.1.3. Ignition source of the ATEX present in the second and third tanks

Both the second and third tanks were identical:

– each tank was equipped with a breathing vent, i.e. an opening that made the inside of the tank communicate with the outside;

– such a vent is essential to ensure the mechanical resistance of the tank, during a filling or an emptying operation. If such a vent were absent, inside the tank there would be, respectively, an overpressure or a vacuum, which would be sufficient for the tank ferrule to collapse;

– the breathing vent may consist of a simple tubing through the roof and open at both ends; it can also be a device combining two valves, a so-called "pressure vacuum relief valve":

- during the filling, the pressure valve opens, allowing the atmosphere contained in the tank's sky to be discharged outside, preventing any overpressure inside it,

- during the emptying, the vacuum valve opens, allowing the outside air to be introduced into the tank, preventing any vacuum inside it;

– moreover, this dual valve device can be equipped with a flame arrester that prevents a flame present outside the tank from igniting an ATEX present near the vent opening. If the breathing vent was free of a flame arrester, the flame could propagate to the ATEX present inside the tank.

Under the conditions of the accident, the second and third tanks were equipped with a vent free of a flame arrester and two possible sources of ignition of the ATEX inside them should be considered:

– during the emptying, the vacuum valve opened, allowing the outside air to be introduced into the tank, preventing a vacuum inside it;

– ATEX may have been ignited by an auto-ignition mechanism: the flame of the outside fire would have heated the outside wall of the ferrule of each tank to a temperature sufficient to initiate the self-ignition of the ATEX, in contact with the inside wall (the auto-ignition temperature AIT of ethanol is 423°C);

– ATEX could also be ignited by the flame present outside the tank, near the breathing vent and that could have propagated through this vent, from the outside to the inside of the tank.

C23.1.3. *Fire resulting from an ignition of formaldehyde by $KMnO_4$*

According to Médard (1979), ethanol is not the only organic molecule with an oxygenated function that ignites spontaneously when it is in contact with $KMnO_4$ crystals. This is particularly the case for certain aldehydes.

This characteristic may be considered as the cause of a fire, which occurred in a poultry farm during a fumigation treatment of the animal's living space.

This treatment involves the reaction of formaldehyde (HCHO) in a flammable aqueous solution at 20% by weight (its flash point FP is about 70°C), with crystals of $KMnO_4$: as soon as the two products come into contact, when the solution is poured on the crystals contained in a vessel, formaldehyde oxidation to formic acid (HCOOH) starts immediately, according to the reaction [C23.1]:

$$2\ KMnO_4 + 3\ HCHO = K_2O + 2\ MnO_2 + 3\ HCOOH \qquad [C23.1]$$

This reaction is accompanied by the following effects:

– the heat release is so intense that the solution is brought to boil in a few seconds;

– a white smoke consisting of a formaldehyde aerosol is generated (it has the desired bactericidal properties and spreads throughout the building);

– the massive release of formaldehyde and water vapors is accompanied by a projection of $KMnO_4$ and MnO_2 pieces around the vessel, at distances up to about 1–2 m, similarly to a Bengal fire.

In the case of the fire to be investigated, about half an hour after the treatment started, the formaldehyde solution was ignited by $KMnO_4$, the fire propagated to straw on the ground and spread out to the entire building that was completely destroyed.

Lessons learned from the investigation

Both accidents reflect a lack of knowledge of the strong oxidation–reduction reaction between $KMnO_4$ and molecules containing alcohol or aldehyde functions.

The reaction with ethanol is sufficiently exothermic to cause its ignition.

The use of this powerful oxidizer in solid form should therefore be prohibited, and only use in diluted aqueous solutions should be acceptable.

Whereas no previous explosion had occurred when alcohol was introduced into a tank already containing "new" crystals, it should be considered that, in the case of the accident, the explosion that occurred was an unusual sequence of operations:

– usually, the amount of alcohol introduced into the tank is sufficient for the crystals to be immerged;

– on the contrary, on the day of the accident, only a limited amount of alcohol was introduced, so that an interface (air/alcohol–vapor/crystals) remained for several dozen minutes;

– the ignition of alcohol, under conditions where this interface exists, has been demonstrated experimentally;

– a fumigation treatment by formic acid vapors, resulting from the reaction between an aqueous solution of formaldehyde and crystals of $KMnO_4$, should be performed only in the total absence of combustible products that could be ignited by the projections resulting from the reaction.

The value of the ethanol flash point is such that ATEX is present in a tank containing ethanol at room temperature. It would be possible to remove this ATEX by using the inerting technique:

– it consists of replacing a part of the air with an inert gas such as nitrogen;

– in order to be totally rigorous, as a result of this replacement, the concentration of oxygen must be reduced to a value not more than the LOC value, i.e. 10% volume for the ethanol/nitrogen couple;

– to be a reliable safety measure, this technique must be implemented in a controlled and rigorous manner; in particular, it is essential that the oxygen content be constantly measured so that nitrogen is automatically introduced into the tank sky if the oxygen content exceeds the LOC value;

– in addition, it assumes that the tank is equipped with special devices, including a pressure vacuum relief valve;

– moreover, if, during storage, alcohol is not stirred or if no mechanical energy is released elsewhere within the tank, it is permissible not to inert the sky of an alcohol storage tank; but it is therefore imperative to equip its breathing vent with a "pressure vacuum relief valve" with a flame arrester, so as to prevent any flame from the outside propagating to the inside of the tank.

Case 24

Explosion Involving Hydrazine

C24.1. Case presentation

The investigation consisted of determining the causes of a phenomenon observed during a bench test of a rocket engine: a rapid increase in the internal pressure of the tank containing one of the propergols powering the engine occurred.

The propergol, subsequently designated by UHH, consists of a mixture of hydrazine hydrate (HH) and asymmetrical dimethyl-hydrazine (UDMH), with a mass proportion (UDMH/HH) = 3. During the hundred seconds of engine operation, the level of the liquid propergol lowers in the tank. Then, in order to maintain its mechanical hold, it is necessary to keep its internal pressure slightly higher than the external one, which is achieved by introducing the waste gases produced by the engine to its sky, to which water vapor is then added at a temperature exceeding 300°C.

To explain the overpressure observed, we suppose that it was the result of an explosive decomposition of the propergol vapors contained in the tank, heated by the waste gases introduced into it.

To evaluate this hypothesis, an experimental search has been carried out in order to:

– reproduce the conditions of this decomposition by reconstituting the mixture of gases and vapors contained in the tank sky in a test vessel;

– study the influence of the initial composition of this mixture, particularly its content in certain constituents of waste gases, as well as its temperature, and its water vapor content.

C24.1.1. *Description of the experimental conditions*

The main components of the test equipment were:

– the test vessel: it had a 1-liter capacity and a cylindrical shape; it was able to sustain an internal pressure of 200 bar and to be uniformly heated up to 350°C;

– it was equipped with all the devices necessary to:

- vacuum it,

- measure at any time the value of its internal pressure, as well as a rapid variation of that pressure which could correspond to an explosion overpressure,

- introduce into it a gas or a mixture of gases under a fixed pressure value,

- introduce into it a fixed amount of a liquid or a mixture of liquids by means of a septum,

- measure the temperature of the gaseous phase,

- locally heat this gaseous phase at a very high temperature, by means of an electric current through a metallic wire, in order to initiate the explosive decomposition of any atmosphere that may give rise to this type of phenomenon.

The tests consisted of:

– introducing different liquids into the vacuumed vessel such as propergols (UDMH, HH or UHH) or water, heated to a given temperature;

– determining the conditions for which a decomposition of the propergol vapors occurs, under the influence of:

- their own temperature,

- the heating at very high temperature of a metallic wire present in the gaseous phase of the vessel,

- the introduction into the vessel of different gas mixtures, such as a sample of waste gases produced by the rocket engine itself, a mixture of N_2–NO, a mixture of N_2–NO_2 or a mixture of N_2–CO_2–H_2.

After some tests, the gaseous phase of the vessel has been picked and its composition has been analyzed.

C24.1.2. *Results*

Depending on the test conditions, particularly the temperature, each of the propergols UDMH, UHH or HH can give rise to a slow decomposition or to an explosive one.

C24.1.2.1. *Slow decomposition*

An analysis of the variation of the internal pressure, resulting from the introduction of 1 cm^3 of one of the propergols into the vacuumed vessel regulated at temperature T_e between 150 and 350°C, has shown that:

– the vaporization of the liquid sample occurs instantaneously and the pressure obtained P_0, between 500 and 1,500 mbar, depends on the nature of the propergol and on the temperature T_e;

– after this vaporization, there is a slow increase in pressure, up to a limit value ($P_0 + \Delta P$), which reflects the decomposition of the propergol vapor, with the formation of gases;

– the average rate of decomposition of each propergol is faster for higher values of T_e;

– for example, at T_e = 350°C, the complete decomposition of HH, UHH and UDMH is obtained after 2, 6 and 26 mn and corresponds to the values of $\Delta P/P_0$ = 0.15, 0.67 and 1.55, respectively.

C24.1.2.2. *Explosive decomposition*

C24.1.2.2.1. Thermal initiation of explosive decomposition

After the introduction of 2 cm^3 of UHH into the vacuumed vessel regulated at 350°C, an analysis of the pressure record has shown that (see Figure C24.1):

– the instantaneous vaporization of this liquid sample produces a pressure P_0 = 1.9 bar;

– 2 s after the vapor has been locally heated by means of a metallic hot wire at the base of the vessel, a peak appears since the pressure increases up to P_e = 3.9 bar and then decreases until it stabilizes at a final value P_f = 3.8 bar.

This peak is due to an explosive decomposition of the vapor of UHH, initiated by the high temperature of the wire.

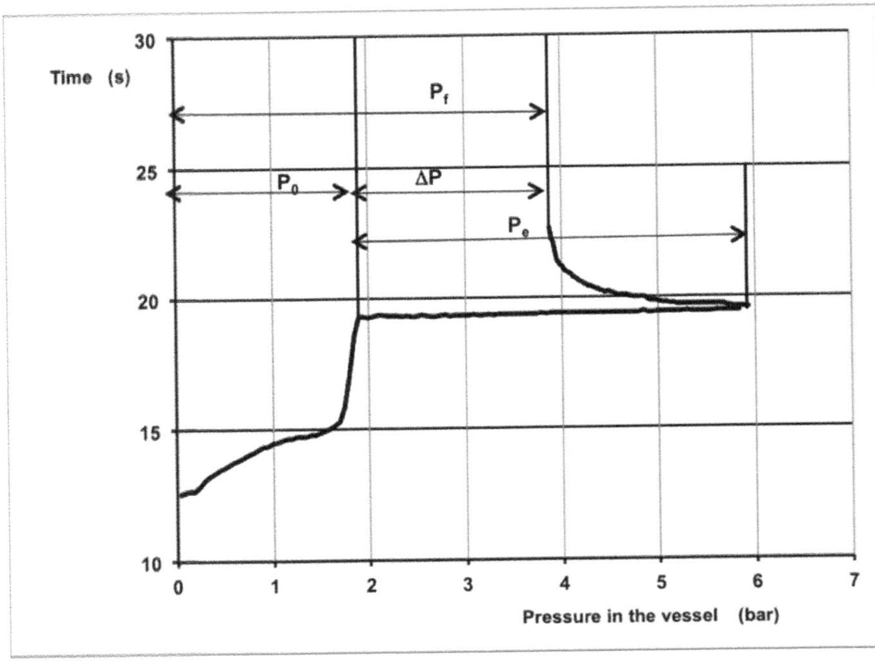

Figure C24.1. *Pressure time record in the vessel during the explosive decomposition of UHH, initiated by a hot metallic wire (thermal initiation)*

UDMH and HH vapors also result in explosive decomposition, with different P_e and P_f values.

In Table C24.1, the values of $\Delta P/P_0$ are summarized, resulting from the slow or explosive decomposition of each propergol, as well as the P_e/P_0 ratio for their explosive decomposition.

Ergol	Slow decomposition	Explosive decomposition by thermal initiation	
	$\Delta P/P_0$	P_e / P_0	$\Delta P/P_0 = (P_f / P_0) - 1$
UDMH	1.55	3.1	1.7
UHH	0.67	2.1	1
HH	0.15	1.1	0.23

Table C24.1. *Comparison of the values of $\Delta P/P_0$ for slow or explosive decomposition of UDMH, UHH and HH propergols*

For each propergol, the different values of $\Delta P/P_0$ show that the nature of the gaseous decomposition products and the composition of their mixture are different, according to the nature of the propergol and whether it is a slow or explosive decomposition.

C24.1.2.2.2. Initiation of explosive decomposition

In an attempt to reproduce the overpressure in the UHH tank during the bench engine test, a test was carried out by successively introducing 1 cm^3 of UHH into the vessel and then, immediately after its vaporization, a sample taken directly from the waste gases produced by the engine.

On the pressure record (see Figure C24.2), the pressure increase corresponding to the vaporization of the propergol (vapor pressure $P_0 = 0.9$ bar) is followed by a new increase to $P_G = 3.2$ bar, corresponding to the partial pressure of the gases of the sample; this second increase is followed immediately by a break of the slope and a peak, with a pressure increase up to $P_e = 2.6$ bar and finally a decrease to a final pressure $P_f = 4.9$ bar. These pressure values correspond to the values of the ratios $P_e/P_0 = 2.9$ and $\Delta P/P_0 = 0.9$, with $\Delta P = P_f - P_G - P_0 = 0.8$ bar.

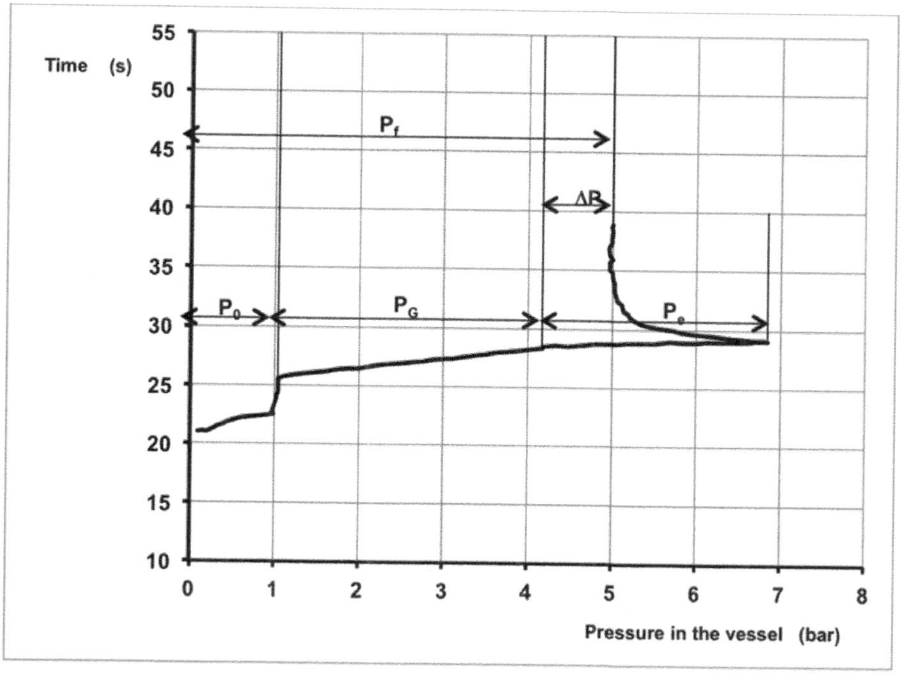

Figure C24.2. *Pressure time record during the explosive decomposition of UHH, initiated by the waste gases (chemical initiation)*

A comparison of the values of $\Delta P/P_0$ corresponding to the thermal and chemical initiation of the decomposition of UHH, i.e. 1 and 0.9, respectively, shows that these values are very close, which is why it is recognized that the initiation mode has no influence on the nature of the decomposition products.

C24.1.2.2.3. Analysis of the decomposition gases of the propergols

An analysis of the gases resulting from the decomposition of each of the three propergols UHH, HH and UDMH has shown that they contain:

– a significant proportion of water vapor (80% volume at 350°C, before its condensation);

– NO and NO_2 oxides at concentrations ranging from 1.2 to 2.9% volume;

– oxygen at concentrations between traces and 3% volume;

– low amounts of the following gases: NH_3, N_2, H_2, CH_4.

Among the flammable components of gases and vapors contained in the propergol tank at the instant of the overpressure, the mass proportion of UDMH and HH vapors is significantly higher than those of NH_3, H_2 and CH_4; N_2 and water vapor are inert gases. Finally, the mixture does not contain oxygen, so that no ATEX is present in the tank and the overpressure in the tank could not be due to the explosion of an ATEX.

C24.1.2.2.4. Initiation of explosive decomposition by waste gases

Tests with different gases or a mixture of gases introduced at temperatures between 200 and 350°C gave the following results:

– there is a delay, up to several tens of seconds, before the introduction of the gas or mixture of gases and the triggering of the explosive decomposition;

– an N_2–NO mixture initiates the explosive decomposition at 350°C, provided that the NO content in the mixture is at least 1.2% volume;

– the same is true for mixtures of N_2–NO_2;

– the decomposition is obtained at a lower temperature when the NO_x concentration of the N_2–NO_x mixture is higher;

– at a given temperature and a given NO_x concentration of the N_2–NO_x mixture, the N_2–NO_2 mixtures initiate the decomposition of propergols more easily than the N_2–NO mixtures;

– like NO_x, oxygen initiates the explosive decomposition of UHH vapors;

– water vapor inhibits the decomposition.

C24.1.3. *Origin of an overpressure in a UHH tank*

The overpressure in the UHH tank was due to the explosive decomposition of the UDMH and HH vapors, initiated by the residual oxides NO and NO_2 contained in the engine waste gases.

These gases are the products of the normal reaction between the propergols and N_2O_4 in liquid state. They are introduced into the UHH tank, but before this introduction, the gases are previously cooled to a temperature of 300°C, and this is achieved by a water injection, so that the gases also contain a high proportion of water vapor (about 80% volume).

Despite this high proportion, the remaining NO_x oxides represent a volumic concentration of a small percentage, sufficient to initiate the explosive decomposition.

Lessons learned from the investigation

A mixture of N_2O_4 and hydrazine derivatives constitutes, in the liquid state, a hypergolic system whose reaction is operated in rocket engines.

The investigation has shown that this same hypergolic reaction also occurs when the propergols are in the vapor state:

– waste gases produced by the engine contain N_2 and a high proportion of water vapor;

– but these gases also contain a few percent volume of nitrogen oxides NO_x;

– despite a high proportion of water vapor, this content is sufficient, at about 300°C, to initiate the decomposition of hydrazine derivatives, present in the sky of the tank in the vapor state;

– in order to prevent this decomposition, it is possible to increase the water vapor content of the sky tank by injecting water into it, which both increases the water vapor content of the sky and lowers its temperature significantly below 300°C.

Case 25

Burst of a Steel Gas Cylinder

C25.1. Case presentation

From the circumstances and effects of the spontaneous burst of a steel cylinder, its origin and the most probable mechanism had to be investigated.

C25.1.1. *Circumstances of the burst*

The bursting of the so-called "target" cylinder occurred during an operation where liquid chlorine contained in a so-called "source" cylinder was transferred to the target cylinder.

C25.1.2. *Effects of the burst*

The burst of the cylinder did not produce any visible thermal effects, but it did produce the two usual mechanical effects for this type of phenomenon:

– the emission of a shock wave;

– the projection of fragments of the cylinder.

For each type of effect, Table C25.1 shows the nature and extent of them, while Figures C25.1–C25.5 show some of them.

For a color version of all figures, in this case study, see www.iste.co.uk/chaineaux/mechanisms.zip.

Type of mechanical effect	Nature of damage	Extent of damage
Production of an aerial pressure wave	Fiber cement roofing and vertical cladding were blown away	The entire roof of the building where the transfer operation was carried out was blown away as well as the vertical cladding (see Figure C25.1)
	Breaking of cinder block walls	The target cylinder was approximately 2 m from the angle of two walls built with cinder blocks which broke between the ground and 1 m high, far from their joints (see Figure C25.2)
Projection of fragments of the target cylinder	Damage caused by numerous fragments (see Figure C25.3) projected around (see Figure C25.5)	Impact of target cylinder fragments on nearby stored cylinders (see Figures C25.4 and C25.5), causing a perforation of some of them

Table C25.1. *Nature and extent of damage*

NOTE.– In the following, the term "burst" is used to reflect the fact that the explosion (or rupture) of the cylinder was accompanied by the projection of fragments:

– the terms "explosion" and "rupture" are therefore more general than the term "burst";

– however, the terms "rupture pressure" and "burst pressure" are used indifferently.

Figure C25.1. *Blowing of the building roof and vertical cladding*

Figure C25.2. *Corner of two cinder block walls near the target cylinder, where some cinder blocks broke, but not their joints*

Figure C25.3. *Some of the numerous fragments of the target cylinder (one was found about 200 m from the center of the burst)*

Figure C25.4. *The source cylinder was punctured by two fragments of the target one*

Figure C25.5. *Set of cylinders about 3 m apart from the target cylinder, some of which having been punctured by its fragments, some of them through*

C25.1.3. *Determination of the mechanism of the burst*

The explosive system involved in the explosion was not obvious nor readily known and, to try to identify it, we applied the method described in Chapter 2:

– the circumstances of the burst do not provide any decisive information, apart from the fact that at this moment, the target cylinder contained some liquid chlorine;

– it is therefore by considering the different energies that were involved in the burst, until producing the effects observed, that it may be possible to determine its origin;

– the burst of the target cylinder resulted from a sequence of phenomena, which have to be identified by estimating the energy released by each of them and to verify the coherence in the hierarchy of these energy values.

Since the rupture of the target cylinder is the central phenomenon, its rupture pressure P_r must be first estimated by means of Lamé's relation, which is presented in section 1.3.2, about the explosive system 3 corresponding to the burst of a hollow body.

In the next section, after providing additional information on the establishment of this relationship, the calculation of P_r is presented.

C25.1.3.1. *Calculation of the rupture pressure P_r of the target cylinder*

Let us consider a closed hollow body, having a cylindrical shape subjected to an increasing internal pressure P and whose rupture pressure is sought to be assessed P_r (the geometric characteristics of the hollow body ends have not to be precisely defined, nor their rupture pressure, which have only to be assumed to be greater than P_r). The characteristics of the cylinder are (see Figure C25.6):

– diameter D (in mm);

– thickness e (in mm);

– internal pressure P (in Pa);

– sigma (σ), the stress of the cylinder material during the pressure increase (in MPa).

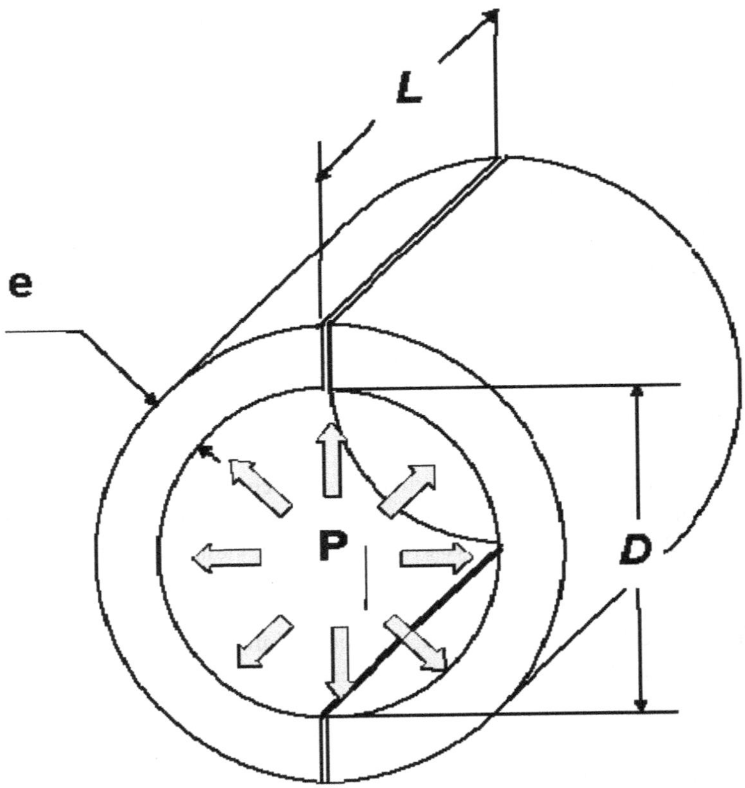

Figure C25.6. *Diagram of the pressure forces to which a cylindrical section (diameter D and thickness e) is subjected, far from its ends, when it is under an internal pressure P)*

During the increase in pressure, a cylindrical section of length L, sufficiently far from the ends of the cylinder so as not to be under the influence of their own mechanical behavior, is subjected to the following different forces:

– a pressure force F_1 that tends to separate the two fictitious half-cylinders that would result from an axial cut of the cylinder (see Figure C25.7);

– a cohesive force of steel F'_1 which opposes F_1;

– a pressure force F_2 that tends to separate the two fictitious sections that would result from a cut of the cylinder in a plane perpendicular to its axis (see Figure C25.8);

– a cohesion force of steel F'_2 which is opposed to F_2.

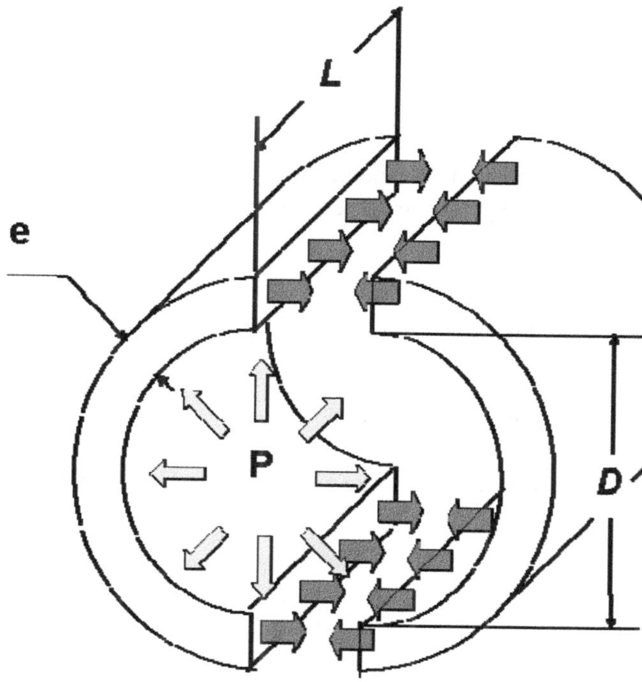

Figure C25.7. *Diagram of the tangential stress σ_{tang} to which a section of the cylinder is subjected, far from its ends, when it is under internal pressure P*

The force F_1 is calculated from the values of P and the pressed surface S_1 (with $S_1 = L.D$):

$$F_1 = P.S_1 = P.L.D$$

The F'_1 force, which tends to maintain the cohesion of the two half-cylinders, is worth:

$$F'_1 = \sigma_{tang}.s_1 = \sigma_{tang}.2e.L$$

where σ_{tang} is the tangential stress in steel (σ_{tang} is homogeneous at a pressure) and where s_1 is the stress surface (with $s_1 = 2e \times L$).

During the increase of P, F_1 is equal to F'_1 and the tangential stress σ_{tang} is therefore written as in relationship [C25.1]:

$$\sigma_{tang} = P.D/2e \qquad [C25.1]$$

The force F_2 is calculated from the values of P and the pressed surface S_2 (with $S2 = \pi \times D^2/4$:

$$F_2 = P.S_2 = P.\pi.D.2/4$$

The force F'_2, which tends to maintain the cohesion of the two halves of the cylinder supposed to be cut by a plane perpendicular to its axis, is worth:

$$F'_2 = \sigma_{long}.\ \pi.D.e$$

where σ_{long} is the longitudinal stress in steel, where s_2 is the stress area (with $s_2 = \pi.D$ and $D > e$).

During the increase of P, F_2 is worth F'_2.

The longitudinal stress σ_{long} is therefore written as in relationship [C25.2]:

$$\sigma_{long} = P.D/4e \qquad [C25.2]$$

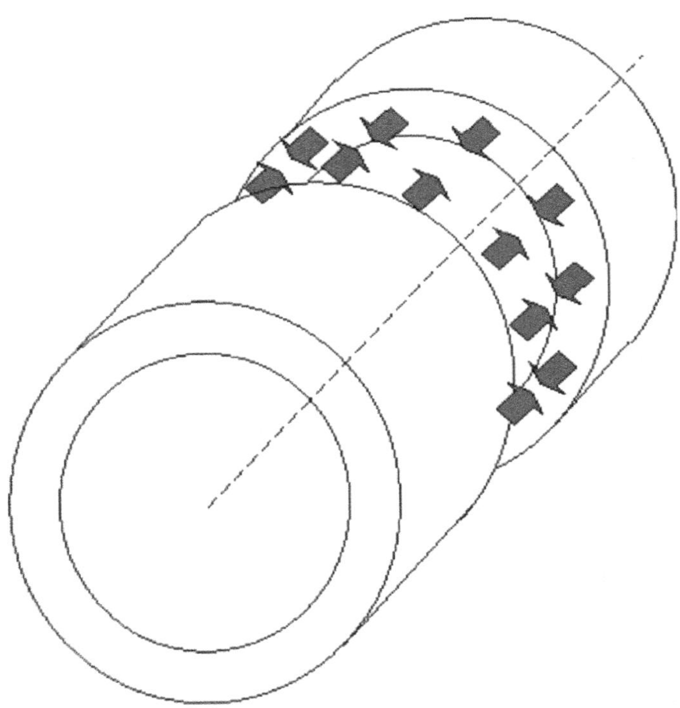

Figure C25.8. *Diagram of the longitudinal σ_{long} stress to which a section of the cylinder is subjected, far from its ends, when it is under internal pressure P*

From the above, it is clear that:

– the longitudinal stress σ_{long} is equal to half the tangential stress σ_{tang};

– the rupture of a cylinder occurs when the tangential stress σ_{tang} reaches the rupture stress σ_r.

This is what we observe: the rupture occurs in a "buttonhole" shape (see Figure C25.9) and the pressure of rupture P_r is given by the Lamé relationship [C25.3] which is identical to [1.35]:

$$P_r = 2e.\sigma r/D \quad [C25.3]$$

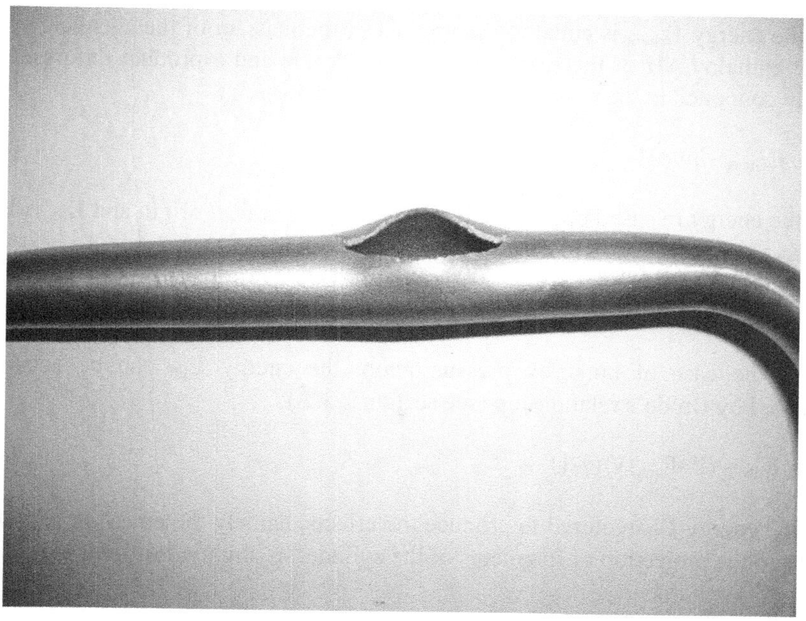

Figure C25.9. *View of a steel pipe, resulting in a "buttonhole" shape rupture under an internal pressure greater than its breaking pressure P_r*

C25.1.3.1.1. Application to the target cylinder

Based on the characteristics of the target cylinder:

– e = 6 mm;

– D = 230 mm;

– σ_r = 1,030 MPa, for the steel grade of the cylinder.

The calculation gives:

$P_r = 530$ bar

C25.1.3.1.2. Sequence of phenomena leading to the observed effects

Table 25.2 gives the sequence of phenomena that could successively occur inside the cylinder before its burst, as well as the energies released.

The differences between the values of the energy released are governed by the fundamental principle of the conservation of energy, corrected for each phenomenon by the existence of a yield less than 1:

– the energy $\mathbf{E_{chem}}$ is equal to the product of the mass **m** of the chemical system by the enthalpy $\mathbf{\Delta H}$ of the reaction between chlorine and a product unforeseeable, already contained in the target cylinder (see section 1.2.2.):

$$\mathbf{E_{chem}} = m\,\Delta H \quad [1.33]$$

– the energy released during the burst $\mathbf{E_{lib}}$ has a chemical origin and $\mathbf{E_{lib}}$ is lower than $\mathbf{E_{chem}}$ because the yield of the chemical reaction is less than 1:

$$\mathbf{E_{lib}} < \mathbf{E_{chem}}$$

– in the case of burst by pressurization, the energy $\mathbf{E_{lib}}$ can be accurately calculated by Brode's relationship (see section 1.3.2.):

$$\mathbf{E_{lib}} = (P_r\text{-}P_{atm})V/(\gamma\text{-}1) \quad [1.34]$$

– the energy $\mathbf{E_{eff}}$ required to produce the effects, namely the emission of a shock wave and the projection of fragments of the cylinder, is slightly less than $\mathbf{E_{lib}}$;

$$\mathbf{E_{eff}} < \mathbf{E_{lib}}$$

Finally, we have:

$$\mathbf{E_{chem}} > \mathbf{E_{lib}} > \mathbf{E_{eff}}$$

In the present case, assuming that the discharged gases are mainly diatomic gases (Cl_2 and HCl, for which $\gamma = 1.4$) and considering the maximum volume of gases before expansion, i.e. V = 40 L, the calculation leads to:

$$\mathbf{E_{lib}} < 1.1 \text{ kg Eq}_{TNT}$$

<table>
<tr><th>Initial phenomenon</th><th>Energy involved</th><th>Effects of the chemical reaction</th><th>Energy involved</th><th>Effects of burst</th><th>Energy involved</th></tr>
<tr>
<td rowspan="2">Chemical reaction involving a mixture of liquid chlorine and one or more unforeseeable reagents contained in the target cylinder (this phenomenon corresponds to an explosive system 2)</td>
<td rowspan="2">The energy released $\mathbf{E_{chem}}$ has a chemical origin
$\mathbf{E_{chim} = m\ \Delta H}$.
m is the mass of the chemical system
ΔH is the enthalpy of the reaction between chlorine and one or more unforeseeable products
Chemical reaction involving a mixture of liquid chlorine and one or more unforeseeable reagents contained in the target cylinder (this phenomenon corresponds to an explosive system 2)</td>
<td>Burst of the target cylinder, as a result of its pressurization by one or more gases produced by the chemical reaction (probably Cl_2 and HCl), until it reaches its breaking pressure $\mathbf{P_r}$.
This phenomenon corresponds to an explosive system 3</td>
<td>The energy released during the burst, $\mathbf{E_{lib}}$, has a physical origin: it comes from the depressurization of the gases produced by the chemical reaction, from the rupture pressure $\mathbf{P_r}$ down to the ambient pressure $\mathbf{P_{atm}}$.
The calculation of $\mathbf{E_{lib}}$ uses Brode's relation:
$\mathbf{E_{lib} = (P_r - P_{atm})V/(\gamma - 1)}$
where:
V is the volume above the liquid in the cylinder,
γ is the ratio of the thermal mass capacities</td>
<td rowspan="2">Emission of a shock wave and projection of fragments of the cylinder</td>
<td rowspan="2">The energy $\mathbf{E_{eff}}$ required to produce the effects, namely the emission of a shock wave and the projection of fragments of the cylinder, is slightly less than $\mathbf{E_{lib}}$
$\mathbf{E_{eff} < E_{lib}}$</td>
</tr>
<tr>
<td>Burst of the target cylinder as a result of the detonation of the explosive system 2 (without prior pressurization)</td>
<td>The energy released during the burst $\mathbf{E_{lib}}$ has a chemical origin. However, $\mathbf{E_{lib}}$ is lower than $\mathbf{E_{chem}}$ because the yield of the chemical reaction is less than 1
$\mathbf{E_{lib} < E_{chem}}$</td>
</tr>
</table>

Table C25.2. *Scheme of the sequence of phenomena that led to the burst of the target cylinder, as well as the energies involved*

Finally, if it is supposed that $\mathbf{E_{chem}}$ is close to $\mathbf{E_{lib}}$, the values of m and ΔH must be such as:

$$\mathbf{E_{chem}} < m\,\Delta H = 1.1 \text{ kg Eq}_{TNT}$$

C25.1.3.1.3. Analysis of the aspect of the cylinder fragments

The bursting of the cylinder produced many fragments (see Figure C25.3); an analysis of the fracture surfaces of the fragments of a hollow body can provide information on the mechanism of its burst.

In the next section, the information that it is possible to draw from this type of analysis is given, in general and in the particular case of the target cylinder.

C25.1.3.1.4. Aspect of fragments resulting from a pneumatic burst

Comparison between "static" and "dynamic" ruptures

The behavior of a hollow body submitted to a pneumatic burst depends on the conditions of its bursting. In order to compare their behavior, the burst of two cylindrical containers (volume 2 m^3) has been carried out.

Both containers had identical geometric and structural characteristics:

– diameter D = 1.25 m;

– height H = 2 m;

– cylindrical shell welded along a generatrix of the cylinder (thickness e = 5 mm);

– 2 GRC bottoms (thickness e' = 6 mm);

– made of carbon steel E24 (σ_r = 400 MPa);

– presence of a 15/21 fitting at half-height.

Each container has been submitted to a burst:

– the first burst resulted from increasing pressure of nitrogen; the pressurization was very slow and could be considered as "static" until the burst, which occurred at rupture pressure P_r;

– the second burst resulted from the rapid increasing pressure produced by the explosion of an ATEX consisting of a stoichiometric propane–air mixture and contained in the container.

Figure C25.10. *View of one container*

The conclusions drawn from these two tests are as follows:

– The first container burst at a rupture pressure $P_r = 35$ bar, a value almost equal to that calculated with Lamé's relationship;

- The rupture of the first container began at the most fragile point of the cylindrical shell, namely point A where one of its generatrixes is tangent to the weld of its tapping (see Figure C25.12); at this point, the shell could not deform as much as the shell itself and the rupture started when the tangential stress σ_{tan} reached the value of the fracture stress σ_r.

- The burst produced three fragments, which all have the characteristics of a ductile fracture, since the thicknesses of the shell and both bottoms have decreased, by 26% and 7%, respectively, while the circumference of the shell at 2 cm from one bottom weld increased by 7%.

- The rupture of the second container also started at point A (see Figure C25.11); two primary fracture fronts developed symmetrically from point A, first in a straight line following the generatrix from A to its intersections B and C with the bottom welds. From there, each front split into two secondary fronts which propagated in opposite directions, three of them strictly following the bottom welds, on both sides of B and C, and the 4th front propagating on a 40 cm length of the bottom before returning to follow its weld. Finally, the four fronts stopped following the welds of the bottoms, and propagated in the shell itself, until both secondary fractures coming from each bottom meet together, thus completing the separation of the three fragments.

- In several locations, the cross-sections of the second container fragments exhibit a brittle fracture surface with a structure where chevrons are embedded in each other, with their points directed towards the point of initiation of the fracture (see Figure C25.11). In this case, the thinning of the shell was much smaller than in the case of the first container.

- Finally, the burst of the second container resulted from the internal explosion of an ATEX and was obtained when the pressure reached a value of about 20% higher than P_r.

- This difference could reflect a general law, namely that, if P_{rdyn} is the rupture pressure obtained under the effect of a rapid increase in the pressure, and if P_{rstat} is the rupture pressure obtained under the effect of a slow increase in pressure, it appears that $P_{rdyn} > P_{rstat}$. However, the difference between P_{rdyn} and P_{rstat} has yet to be quantified.

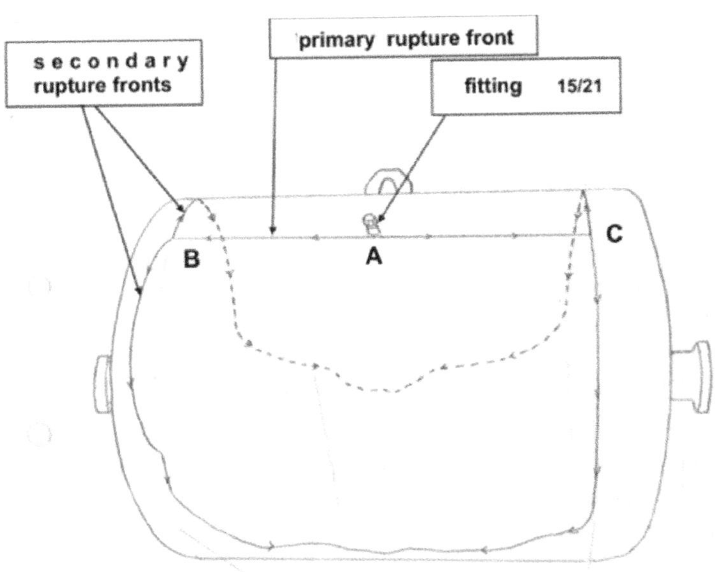

Figure C25.11. *Diagram of the propagation of the different rupture fronts*

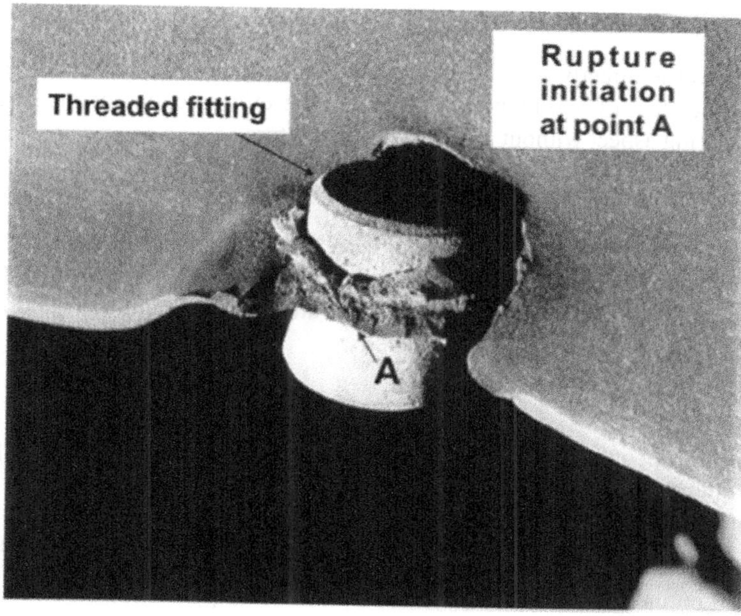

Figure C25.12. *Location of point A, at the point of tangency between the cylinder generatrix and the weld of the tapping*

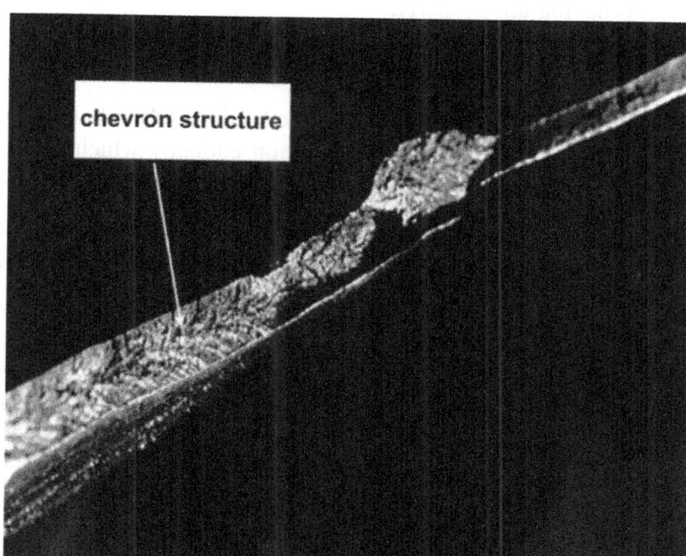

Figure C25.13. *Aspect of the cross-section of a fragment of the second container, exhibiting a brittle fracture having a chevron structure*

Case of steel rupture under a detonation regime

When steel is directly in contact with a detonating explosive material, as is the case with the steel of a shrapnel for example, the cross-sections of the fragments exhibit a tearing aspect without any organized structure.

Application to the fragments of the target cylinder

Examination of the cross-sectional surfaces of the fragments (see Figure C25.3) did not show any chevron characteristic of a burst by pressurization. On the contrary, the burst produced numerous and rather small fragments exhibiting a tearing aspect without organized structures.

Attempt to identify the explosive system

It can be supposed that the burst resulted from the detonation of an explosive system contained in the target cylinder as a mixture of liquid chlorine and one or more unforeseeable products. Then, the products that are likely to spontaneously give rise to an exothermic chemical reaction have been investigated as follows:

– based on thermochemical considerations, the enthalpy of the chemical reaction between chlorine and particular oxidizable products has been calculated, to identify those for which this reaction is highly exothermic;

– then, a brief experimental campaign has been carried out, aiming to identify explosive systems having this property.

Analysis of the thermochemical system (Cl_2 + C_2H_4O)

Among the unforeseeable products considered, ethylene oxide C_2H_4O, a liquid product boiling at 20°C with a negative formation enthalpy which gives it a high reactivity with powerful oxidants such as Cl_2:

– the reaction [C25.4] between chlorine and ethylene oxide involves one mole of ethylene oxide for one mole of Cl_2 and, for two moles of reagents, it produces three moles of gas (carbon remains solid):

$$C_2H_4O + Cl_2 = 2\,C + H_2O + 2\,HCl \qquad [C25.4]$$

– its enthalpy ΔH is obtained from the enthalpies of formation of reagents and products, i.e.:

$$\Delta H = -\,2.9 \text{ kJ/g for the stoichiometric system } (C_2H_4O + Cl_2)$$

– this value characterizes a moderate exothermic reaction, since it represents 62% of the decomposition energy of TNT, i.e. 4.69 kJ/g;

– the mass m of the explosive system which was necessary for the pressure of the gases released by the reaction in the volume of the target cylinder, i.e. V = 40 dm^3, to reach the value of its rupture pressure P_r = 530 bar under stoichiometric conditions. The value of m is obtained:

- from adiabatic flame temperature T_f of the reaction, calculated from the coefficients of the reaction and from the thermal mass capacities of the reagents and products, i.e. T_f= 2264 K,

- from the number of moles of the reagents, which is related to P_r, T_f and V by the state equation of the perfect gases, as well as the number of moles of the gases produced by the reaction, the ratio of them being 3/2,

- from the molecular mass of the stoichiometric system, M = 0.115 kg/mole.

The calculation leads to m = 20 kg, which represents a TNT equivalent of 12 kg. This value of m is relatively low, compared to the maximum mass of the products contained in the target cylinder, i.e. 50 kg. Therefore, it can be admitted that a sufficient mass of the explosive system combining chlorine and ethylene oxide could have been formed in the cylinder, so that its explosion could have developed a sufficient pressure to cause the cylinder bursting.

C25.1.4. *Conclusions of the investigation*

At this stage of the investigation, the conclusions are the following ones:

– the previous calculations correspond to a system assumed to be stoichiometric. However, it is possible to say that it was certainly not, at least because the reagents did not have time to mix; the system mass needed to burst was therefore greater than 20 kg;

– however, in case of a detonation of the system and as discussed above, the burst could possibly result from the application of mechanical forces sustained by the internal wall of the cylinder, which was directly in contact with the explosive system, before the pressure reaches P_r; the mass of the system could therefore have been less than 20 kg;

– in the end, it is not possible to precisely estimate the mass of the non-stoichiometric chemical system that exploded, nor the value of the chemical energy E_{chem} released (see Table C25.2).

A spontaneous initiation of an explosive reaction has been observed when a mass of 20 g of ethylene oxide was put in contact at room temperature with a mass of 26 g of chlorine, initially contained in a 0.5 L test vessel. This explosion developed a pressure which has not been measured but which could be estimated at about 100 bar. Then, it can be concluded that:

– ethylene oxide is a product capable of a violent and spontaneous reaction with chlorine;

– for a sufficient mass of such an explosive system, this reaction could have led to the target cylinder bursting;

– however, there is no evidence or circumstance of the accident that would favor this product as the unforeseeable one.

Lessons learned from the investigation

– The process of establishing the mechanism of an accidental explosion of which the explosive system is not known at the outset, as may be the case when a hollow body bursts, first, an estimate of the energy required to produce the observed effects;

– the burst of the hollow body may have resulted from a sequence of several consecutive phenomena and in this case, the mechanism to be established must take into account both:

- the existence of a hierarchy between the energy released by a phenomenon and the energy necessary for the occurrence of the following one,

- the fundamental principle of energy conservation, tempered by the fact that any transfer of energy is affected by a yield less than 1;

– an analysis of the surface of the fragments resulting from the bursting of a hollow steel body can provide information on the burst mechanism:

- the rupture pressure P_r at which a cylindrical hollow body bursts during a slow increase of its pressure can be estimated relatively accurately by means of the Lamé relationship; the fragments of the hollow body show the characteristics of a ductile fracture, with a significant decrease in the thickness of its wall,

- the pressure at which a cylindrical hollow body bursts during a rapid increase in pressure, under the effect of an internal ATEX explosion, is significantly greater than P_r; the edges of the fragments of the hollow body can then show a chevron structure characteristic of a brittle fracture with a slight decrease in the thickness of its wall,

- finally, a mechanism for detonating a solid or liquid explosive system contained in the hollow body must be retained when the fragments are numerous, with cross-sections having a shredded appearance, without an organized structure;

– the energy released when a hollow body bursts can be estimated relatively accurately using the Brode relationship, which needs to know the volume of the hollow body and the P_r value;

– if the burst circumstance analysis does not enable us to identify the nature of the explosive system, it should be considered that it consists of a mixture of products likely to give rise to a violent chemical reaction, having the following characteristics:

- it starts spontaneously as soon as the components of the system are brought into contact (so-called "hypergolic" system),

- the estimation of its enthalpy, based on the thermochemical characteristics of its constituents, leads to a very strongly exothermic reaction,

- in view of the estimated value of this enthalpy, the mass of the explosive system necessary for the release of energy at least equal to that necessary for the production of the observed effects must be compatible with that which might be present in the hollow body, in particular in view of its volume;

– once a theoretical explosive system has been identified, the only way to verify that it meets the above conditions is through experimental validation.

Case 26

Explosion in a Foundry of Steel Waste

C26.1. Case presentation

An explosion occurred during the operation of an arc furnace, which was used to melt steel waste at 1,400°C, and its mechanism had to be investigated.

C26.1.1. *Description of the facility*

The process involves melting the recovered metal (mainly steel) in a furnace, and then tipping the furnace to cast the molten metal in the mold.

During melting, the slag formed on the surface of the metal must be removed by successive pours into containers of about 1 m^3 capacity, placed under the furnace, in the slag pit.

C26.1.2. *Circumstances of the explosion*

The explosion occurred when the slag pit, where was a container full of slag being cooled, has been abnormally filled with water in such amount that the container has been totally submerged.

C26.1.3. *Effects of explosion*

The furnace was overturned after being removed from its tipping bearings, while its mechanical support suffered very serious disorders (see Figure C26.1). This

For a color version of all figures, in this case study, see www.iste.co.uk/chaineaux/mechanisms.zip.

displacement means that the furnace has been submitted to a vertical force, sufficient to lift it and, on the basis of the diameter of the furnace, i.e. 4 m, and the mass of the furnace, i.e. 20 t, the minimum pressure necessary to exert this force can be estimated to 160 mbar.

Figure C26.1. *View of the overturned furnace and the disorder of its support*

Several pieces such as debris from the support of the furnace or deck components that covered the slag pit (the heaviest debris weighed 635 kg) were thrown vertically and, after passing through the roof of the building housing the facility, fell outside at 90 m from the furnace.

Many of the building's windows as well as the roof and wall cladding sheets were blown.

C26.1.4. *Estimation of E_{eff} based on damage analysis*

The analysis of the damage produced in the far field (breaking of the cladding, projection of debris) enables us to estimate the maximum value of the energy E_{eff} which was necessary for the production of this damage, i.e. 50 kg Eq_{TNT}.

C26.1.5. *Determination of the explosion mechanism*

The following circumstances explain the mechanism by which water was brought into contact with the molten slag, thus causing instantaneous vaporization of a large amount of water:

– the tank in the pit contained 3 T of slag which, because of its low thermal conductivity and the duration of its cooling in air, had a solidified external surface but still a liquid core, at an average temperature that can be estimated at 1,100°C;

– water gradually filled the slag pit and eventually submerged the container;

– thermal stresses have developed in the solidified mass of the slag and resulted in its abrupt cracking, which allowed water to penetrate to the core of the slag and come into contact with a large proportion of its liquid fraction;

– the hypothesis that the explosion involved the slag in the container is supported by the fact that after the explosion, the container was completely empty of slag.

The explosion involved an explosive system 4, which is characterized by the release of the energy E_{lib}, the maximum value of which is calculated using the relationship [C26.1], from the following parameters (it is identical to the [1.36] one):

$$E_{lib} = m\,(c_p\,\Delta T + \Delta Hv) = 2742 \text{ MJ} = 585 \text{ kg Eq}_{TNT} \qquad \text{[C26.1]}$$

where:

– m is the mass of water likely to vaporize, estimated at 400 kg;

– ΔT is the temperature difference between water and molten slag, estimated at 1,100°C;

– c_p is the thermal mass capacity of water, i.e. 4,190 J/kg.K;

– ΔHv is the water vaporization enthalpy, i.e. 2,257 kJ/kg.

C26.1.6. *Yield of the explosion*

Although, in this case, the heat transfer did not involve a liquid and a solid, but rather two liquids, it was carried out under conditions which, for a system 4 explosion, are very unfavorable and explain that, for an estimated maximum value of E_{eff}, i.e. 50 kg Eq_{TNT}, the yield ρ was much less than 1:

$$\rho = E_{eff}/E_{lib} << 1$$

Lessons learned from the investigation

Although the yield of a system 4 explosion is always much less than 1, this phenomenon causes damage that is all the more important as:

– the masses of the hot body and the cold liquid are greater (several hundreds of kg);

– the initial contact surface between the hot body and the liquid is greater;

– the hot body is itself in a molten liquid state, at a temperature greater than 1,000°C.

Case 27

Explosion in the Boiler of a Household Waste Incinerator

C27.1. Case presentation

The investigation consisted of determining the causes of an explosion in the boiler of a household waste (HW) incineration plant.

C27.1.1. *Facility description*

The scheme of an HW incineration plant is given in Figure C27.1.

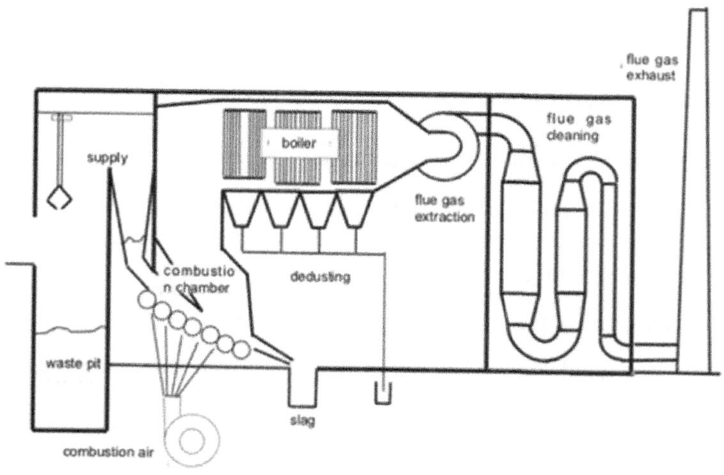

Figure C27.1. *HW incineration plant*

It consists of the following main elements:

– The combustion chamber: fed from the waste pit, it consists at its base of a set of parallel rollers through which the combustion air is blown and which creates a sloping space, allowing the evacuation by gravity of the ashes and slags. The fumes produced are driven up the chamber, the wall of which is lined with a network of steel tubes, where water circulates and which are covered with a refractory material to preserve them from too high temperatures.

– The boiler: after a vertical path towards the top of the combustion chamber, the fumes are driven towards a boiler with water and steam tubes, arranged horizontally, which consists of a welded sheet metal block of about 800 m^3 volume. Inside this volume are arranged a set of parallel tubes in which either water or steam circulates (at 260°C under 40 bar, in the overheating part) and to this set is associated a reserve of water, contained in a tank of several m^3, whose sky contains air under the same pressure.

– The fume treatment unit: downstream of the boiler, the fumes are dedusted and then sucked by an exhaust fan located at the boiler outside, into a cleaning unit before being sent to the chimney.

C27.1.2. *Circumstances of the explosion*

The incinerator was operating under nominal conditions, which correspond to the introduction of HW at a flow rate of 8 T/h and a combustion chamber temperature slightly above 1,000°C when, without a signal alarm nor warning sign, an explosion occurred, causing extensive damage to the boiler.

C27.1.3. *Description of damage*

The boiler suffered the most severe damage:

– the block-shaped boiler casing, consisting of welded plates, was subject to the effects of internal overpressure, which severely deformed the casing, resulting in tearing of welds in several areas;

– leak orifices have appeared on several boiler tubes:

- on the one hand, one of the tubes in the vault of the combustion chamber opened on a generatrix, the surface of the opening being only slightly less than the right section of the tube. Such a rupture, known as a "buttonhole" (see Figure C25.9), occurs as soon as, under the effect of the internal pressure of the tube, the longitudinal rupture stress of the steel is reached,

- on the other hand, several welds ruptured at the connections between several tubes of the boiler manifold, revealing as many leak orifices of smaller cross-section as that of the vault tube opening.

On the contrary, upstream of the boiler, the combustion chamber was not damaged and downstream, the exhaust fan casing was only slightly deformed.

C27.1.4. *Determination of the explosion mechanism*

Before establishing the explosion mechanism, it is relevant to present the general case of an explosion occurring in a combustion plant consisting of a combustion chamber fueled with gaseous or solid fuel and associated with a boiler.

C27.1.4.1. *Possible scenarios*

An explosion occurring in such a facility may a priori involve either the combustible product itself or an element of this facility.

C27.1.4.1.1. Cause related to the nature of the products introduced into the facility

In the particular case of an HW incinerator, the waste to be incinerated consists of a mixture of solid products, the nature of which is very variable and difficult to control, and it may happen that, among these products, there are certain objects, relatively strange to normal HW:

– It may be objects containing substances that may explode at the high temperature of the combustion chamber (chemical products, firework components, ammunition, etc.), and which correspond to a type 2 explosive system.

– It may also be closed hollow bodies (empty aerosol can or containing a residual propellant gas, cylinder of any material and volume, up to several liters). The high temperature of the combustion chamber increases the pressure inside the hollow body until it bursts. The damage produced results from both the overpressure wave and the effect of the burst. This event, which corresponds to the explosive system 3, also causes the contents of the hollow body to be dispersed, which may be followed by its combustion, if it is a flammable product.

– All these phenomena generate both overpressure and bursting, which can cause more or less damage to the combustion chamber, depending on the mass of the explosive material or the volume and pressure of the hollow body bursting.

C27.1.4.1.2. Cause of failure of a facility component

The combustion chamber and the boiler are the components of the facility whose failure may cause an explosion, according to one of the following mechanisms:

– In a gas-fired combustion plant, if the gas entering the combustion chamber does not ignite in the combustion chamber, either because the chamber is not simultaneously supplied by a sufficient flow of air or because the pilot light remains extinguished while a non-ignited ATEX flows at the main burner and accumulates there, which constitutes an explosive system 1. The addition of a source of ignition, for example, during the too late ignition of the pilot light, initiates its explosion, by producing mechanical damage all the more as the volume of the ATEX is greater.

– In case of too much heat stress, one of the tubes of the boiler or the ones lining the wall of the combustion chamber can be punctured: the water which then invades the boiler or the combustion chamber vaporizes massively in contact with their very hot wall. This is an explosive system 4, since the water steam produced instantly develops an overpressure that can be sufficient to produce significant damage, both in the boiler and in the combustion chamber.

– The boiler circuit has a water storage tank that contains pressurized air up to several tens of bars. It is a pressure vessel constituting an explosive system 3, the mechanical failure of which may lead to its bursting.

C27.1.4.2. *Explosion mechanism in the HW Incinerator*

The damaged incinerator being a solid fuel combustion facility associated with a boiler, the explosion which occurred there certainly belongs to one of the four types of explosive systems.

To identify the explosion mechanism, the hypothesis of each type of explosive system has been evaluated.

C27.1.4.2.1. ATEX explosion hypothesis (system 1)

The high temperature of the combustion chamber is sufficient to ensure the combustion of any combustible material, gaseous or solid, introduced into the combustion chamber, and it is therefore not possible for an ATEX to form and accumulate before its ignition.

In the present case, this hypothesis can be rejected, since the explosion occurred while the plant was operating normally and, according to an analysis of the concentration of various gases, including CO contained in the fumes, their composition corresponded to the nominal operating conditions of the incinerator.

C27.1.4.2.2. Hypothesis of the decomposition of a solid substance (system 2)

The combustion chamber is not an element whose mechanical strength is significantly greater than the boiler and the fact that the damage was produced only in the boiler makes it possible to reject this hypothesis.

C27.1.4.2.3. Hypothesis of a hollow body burst (system 3)

The explosion did not give rise to the bursting of any hollow bodies, either in the water tank or in the combustion chamber. Furthermore, in the latter case, it is likely that damage would also have occurred in the combustion chamber and this hypothesis can also be excluded.

C27.1.4.2.4. Hypothesis of a sudden water vaporization (system 4)

From the opening of the vault tube or the drain orifices of the collector, water has invaded the boiler and its sudden vaporization has developed an overpressure. This is why this hypothesis has been completely evaluated, because the damage suffered by the boiler actually resulted from this vaporization.

In addition to having to validate this hypothesis, the question then arose as to whether the opening of the vault tube was the initiating event and whether the rupture of the manifold welds was the consequence, or whether the reverse chronology occurred.

C27.1.4.3. *Validation of the hypothesis of a sudden vaporization of water*

To validate the hypothesis of a sudden vaporization of water, the following estimations or verifications had to be carried out:

– estimation of the mass flow rate of the water leak;

– estimation of the volume flow of steam produced;

– verification that the steam flow rate was greater than the exhaust fan flow rate, and that the difference was sufficient to explain the development of an overpressure in the boiler until it reached its breaking pressure.

C27.1.4.3.1. Estimated water leak and steam flow rates

The mass flow rate of the water leak and the volume flow rate of the steam produced depend on the pressure in the boiler system and on the section of the leak opening.

In the overheat boiler part, the steam was at 260°C, corresponding to a saturating pressure equal to 47 bar.

Table C27.1 presents:

– the characteristics of the two considered leak openings, on the one hand, the opening of the vault tube and, on the other hand, the leak opening resulting from torn welds at the boiler manifold;

– the water leak mass flow rate and steam volume flow rate values for each type of leak opening.

Leak orifice characteristics		**Mass flow rate of water leak** kg/s	**Volume flow rate of produced steam** Nm3/s
Origin	**Surface** cm^2		
Vault tube opening (diameter 100 mm)	Buttonhole opening with an estimated surface of 50 cm^2	60	100
Torn welds between boiler manifold tubes	Surface equivalent to an 8 cm^2 rectangle (40 cm long and 2 mm wide)	10	15

Table C27.1. *Mass flow rate values of the water leak and volume flow rate of the steam produced, corresponding to the two types of leak opening envisaged*

The values of the volume flow rates of produced steam in Table C27.1 should be considered as maximum values, taken at the beginning of the leak, when the pressure and the temperature in the overheat boiler part are maximum.

C27.1.4.3.2. Fume extraction rate

The nominal exhaust fan flow rate was 15 Nm^3/s, i.e.:

– more than six times less than the steam volume flow rate corresponding to the vault tube opening, which is worth 100 Nm^3/s;

– as much as the steam volume flow rate corresponding to the rupture of the manifold welds.

In the first case, the situation would have necessarily generated a significant overpressure in the boiler casing.

In the second case, it can be assumed that the overpressure developed in the boiler would have been much lower.

C27.1.4.3.3. Behavior of the boiler under internal overpressure

According to its parallelepiped shape and its constitution (flat faces connected at right angles by welding), the overpressure necessary to cause permanent deformations of its walls, or even the rupture of certain welds, can be estimated at 25 mbar.

A comparison of this value with that, estimated at several bars, which could have been developed in the boiler casing, assumed to be sufficiently resistant for the leak cases envisaged above, validates the hypothesis of an explosion resulting from a sudden vaporization of water in the boiler.

In addition, it is the opening of the vault tube as the initiating event of the explosion that must be preferred for the following reasons:

– the mass flow rate of the corresponding water leak, estimated at 60 kg/s, corresponds to the same order of magnitude as the flow rate at which the water tank was emptied as a result of the explosion;

– the absence of the refractory material, which covered the tubes of the vault to ensure their thermal protection, has been observed after the explosion and it is likely that this absence has led to the rupture of the tube;

– indeed, it entrained a strong increase of the temperature of the tube wall; simultaneously, the tangential rupture stress σ_r of the steel has greatly decreased, as its internal pressure P_r and the tube ruptured when P_r reached a close value to 47 bar, equal to this internal pressure.

C27.1.5. *Protection of the boiler against the effects of an explosion*

The damage to the boiler is due to the fact that the overpressure, resulting from the rupture of a water tube, has reached its rupture pressure P_r.

As much as it may be difficult to prevent the rupture of a water tube, in circumstances similar to those of the explosion, as much as it would have been possible to limit its consequences by implementing the principle of vent protection.

C27.1.5.1. *Explosion vent protection principle*

An explosion vent protection is a commonly used means of protecting an enclosure from the effects of an internal ATEX explosion, whether the fuel is a gas or a dust.

It consists of placing on the wall of the enclosure one or more explosion vents, which are surface elements whose opening pressure P_{stat} is low and in any case significantly lower than the rupture pressure of the enclosure P_{red}.

The total surface of these vents shall be calculated based on:

– the volume V of the enclosure;

– the rupture pressure P_r of the enclosure;

– the opening pressure P_{stat} of the vent or vents;

– the maximum rate of pressure rise $(dp/dt)_{max}$.

The rupture pressure P_r of the enclosure is accessible from its volume V, from considerations relating to its shape and construction characteristics, and finally to the mechanical strength properties of the materials which it is made of; in general, the volumes of the parallelepiped shape are characterized by relatively low P_r values, not exceeding 100 mbar, much lower in any case than those relating to cylindrical shapes.

The P_{stat} value is to be fixed according to P_r, but it is always preferable that it is as low as possible, or in the order of 50 mbar, without, however, risking untimely openings of the vents, depending on the variations in the operating pressure of the process.

When the explosion is initiated, the vent opens at the P_{stat} pressure, but the pressure continues to increase to a P_{red} value, known as "reduced pressure", and it is P_{red} that must remain below P_r.

Finally, for an explosion occurring in a closed explosion-resistant enclosure, the maximum rate of pressure rise $(dp/dt)_{max}$, which is expressed in bar.m/s, characterizes the rate of combustion of the fuel concerned:

– $(dp/dt)_{max}$ is defined as the value of tg β, where β is the angle between the tangent to the inflexion point of the curve P = f(t), with the horizontal (the curve obtained for a stoichiometric methane–air mixture, ignited in the center of a sphere of 2 m^3 and the β angle are given in Figure 1.5);

– it results that the higher the burning rate of a fuel, the higher the $(dp/dt)_{max}$ is.

In addition, for a given fuel at a given concentration, $(dp/dt)_{max}$ depends on the volume V of the enclosure, according to the following relation:

$$(dp/dt)_{max} \cdot V^{1/3} = \text{Constant} \qquad [C27.1]$$

This leads to define the characteristic K_G of a gas, which corresponds to the value of $(dp/dt)_{max}$ measured for an explosion in a spherical 1 m^3 enclosure.

The NFPA Guide 68 "Venting of deflagrations" (NFPA 2007) provides the calculation model that, based on P_{stat}, P_{red} and K_G values, defines the vent surface required to protect a V-enclosure.

Note that the vent protection principle is identical for a dust explosion; the parameter equivalent to K_G is named K_{st} and corresponds to the value of $(dp/dt)_{max}$ measured during a dust explosion in a spherical 1 m^3 enclosure.

C27.1.5.2. *Application to the boiler incinerator*

In the case of the incinerator boiler, taking into account its internal volume, i.e. 800 m^3 and the estimated volume flow rate of the steam produced, i.e. 100 Nm^3/s (Table C27.1), the maximum rate of pressure rise that would result from the sudden vaporization of water can be calculated between 0.1 and 0.2 bar/s.

Such a value corresponds to the one that would be measured during the fictitious explosion of an ATEX, in the same volume V = 800 m^3 and involving a fuel whose value of K_G would be worth 1–2 bar m/s.

This value is relatively low and in any case lower than the K_G of ammonia gas, i.e. K_G = 10 bar.m/s, which is the gas with the lowest K_G.

As a result, the NFPA method is suitable for defining the vent surface needed to protect a boiler from overpressure phenomena having a kinetics at most equal to that of the explosion of an air–ammonia ATEX.

In practice, it would be possible to apply the NFPA method, using the K_G value of ammonia, to size the vents to be installed on the boiler, in order to protect it from the overpressure phenomenon following the rupture of a water tube.

Note, however, that this conclusion would be different for any situation where the pressure rise in the boiler would be greater than 0.2 bar/s (which would happen, for example, for a water leak flow rate greater than 60 kg/s, or for a boiler volume less than 800 m^3).

Lessons learned from investigation

During the operation of a combustion plant associated with a boiler, a water tube presents the risk of a rupture.

In such a scenario, the water discharged into the boiler would vaporize massively in contact with its hot walls, which corresponds to the explosive system 4.

Despite the presence of the fume extraction fan, the steam production flow rate may be sufficient for the overpressure developed in the boiler to reach its breaking pressure.

To take this risk into account and limit the damage, the boiler could be protected by an explosion vent: the vent surface to be installed should correspond to the recommendations of the NFPA guide (NFPA 2007), assuming to consider that the rate of pressure rise in the boiler would be of the same order of magnitude as that resulting from the explosion of an ATEX consisting of a weakly explosive gas such as ammonia whose K_G characteristic is worth 10 bar.m.s^{-1}.

Conclusion

The risk of explosion is present everywhere in human life, not only in domestic situations but also and particularly in industrial activities, where all sectors are concerned.

Every explosion is characterized by the release of a large amount of energy E_{lib} in a very short time and by the production of damage both human and material, the nature and intensity of which being able to be considered as the "signature" of the explosion.

The investigation of an explosion consists of identifying all the elements of its causes and to establishing its mechanism, according to its circumstances and to the damage it has caused. Of course, these elements are an essential step in understanding the phenomenon and the conditions of its occurrence.

In writing this book, I hoped to transmit the experience I acquired in this investigation activity and I chose to present 27 case studies that I selected to describe the broadest spectrum of possible mechanisms.

A first result of this work led me to consider that any explosion involves an explosive system that necessarily belongs to one of the following four types:

– type 1: the explosive system is an explosive atmosphere (ATEX);

– type 2: the explosive system consists either of a mixture of substances which could give rise to a highly exothermic chemical reaction or of a single substance which could break down in an explosive way;

– type 3: the explosive system is a hollow body which breaks and bursts under the pressure of its contents, initially in the state of a gas or a compressed vapor;

– type 4: the explosive system consists of a hot body that comes into contact with a cold liquid and causes its sudden vaporization.

To establish the mechanism of an explosion, the following stages must be achieved:

– Rigorously identify all the circumstances of the explosion, in order to deduce, if possible, the explosive system involved. For 7 upon 27 cases, a short experimental campaign has been carried out and the results obtained have been very useful to validate a hypothesis made about the nature of the explosive system or the conditions of its explosion.

– Estimate the value of E_{lib} from the nature and intensity of damage. Some simple laws or models of physics and chemistry allow this estimation: for example, an explosion occurring in the open air generates a pressure wave and, from a mapping of the damage caused by that wave at the distance R from the center of the explosion, the TNT equivalent model can be used to determine the mass (m) of the explosive system 1, 2 or 4, or the volume and burst pressure of the hollow body for system 3.

– From E_{lib} and the nature of the explosive system, deduce the quantity of the explosive system involved, and verify that it is compatible with the one which can be deduced from the circumstances of the explosion.

– Finally, it is generally possible to identify the main elements of the explosion mechanism, since the formation of the explosive system until the initiation of the explosion and the production of the damage.

In the case of an explosion resulting from a flammable gas leak, data on the flow rate and duration of the leak are often sufficient to determine the mass of ATEX involved and to deduce its source and the conditions of its formation.

It is rare that all the causal elements that allowed the explosive system to form and then explode can be determined (e.g. the source of inflammation of an ATEX may remain unknown) and it is therefore necessary to remain humble and accept that certain aspects of the mechanism remain uncertain or unexplained.

For each case, the lessons learned from investigation were given for two purposes:

– bring together each particularity of an explosive system, the process of the explosion, its circumstances and its effects;

– define the measures which are able to prevent a new explosion or to protect against its effects.

Finally, the activity of explosion investigation is one part of a comprehensive approach to enriching the feedback of accident experience, which is an essential step in defining the means to be used to prevent explosions.

References

Allen, C.B. and Janz, C.J. (1980). Molten salts safety and hazards an annnotated bibliography. *J Hazard Mater*, 4, 145–175.

BARPI (2008). Explosion de gaz dans les unités de craquage catalytique et gas plant d'une raffinerie. Aria sheet no. 3969, Bureau d'Analyse des Risques et Pollutions Industriels.

Chaineaux, J., Janes, A., Sallé, B., Petit, J.M. (2009). Conditions de formation d'une atmosphère explosive lors de la mise en œuvre d'un liquide inflammable. *INRS, Santé et sécurité du travail*, 216, ND 2313, 23–31.

Coward, H.F. and Jones, G.W. (1952). Limits of flammability of gases and vapors. *USBM Bulletin*, B503.

Department of the Army, the Navy and the Air Force (1990). Structures to resist the effects of accidental explosions. Technical Manual, TM 5-1300/NAVFAC P-397/AFR, 88–22.

ExplPress (2006). Safekinex EU project (Projet Safekinex: SAFe and Efficient hydrocarbon oxidation processes by KINetics and Explosion eXpertise). Software, ExplPress.

Frelov, J.E., Mal'tseva, A.S., Baratov, A.N., Rozlovskii, A.J. (1977). Certain laws governing the self-combustion of mixtures of chlorine and hydrogen. *Khim Prom*, 7(5), 530–532.

Gmelin, L. and Meyer, R.J. (eds) (1930). Gmelin Handsbuch der Anorganischen Chemie. System-Nummer 59: Eisen. Teil B, 239.

Harris, R.J. (1983). The investigation and control of gas explosions in buildings and heating plants. *British Gas Corporation*, 14, 194.

Ineris (2005). DRA 35, Omega 11. Méthodes pour l'évaluation et la prévention des risques accidentels – Connaissance des phénomènes d'auto-échauffement des solides combustibles. Report, Ineris.

INRS (2019). La méthode de l'arbre des causes – L'analyse de l'accident du travail, ED 6163, 24. Report, INRS.

Lannoy, A. (1984). Analyse des explosions air-hydrocarbure en milieu libre. Étude déterministe et probabiliste du scénario d'accident – Prévision des effets de surpression. *Bulletin de la Direction des études et recherches, Électricité de France, série A*.

Médard, L. (1999). *Les explosifs occasionnels, Monographies, propriétés*, 2nd edition. Technique et Documentation, Paris.

Molname, M., Schlender, T., Schröder, V. (2003). Caractéristiques de sécurité – Explosibilité des mélanges de gaz. *BAM*.

Norme EN 14491 (2012). Systèmes de protection par évent contre les explosions de poussières.

Norme EN 14494 (2007). Systèmes de protection par évent contre les explosions de gaz.

Norme NF-EN 1127-1 (2011). Atmosphères explosives – Prévention de l'explosion et protection contre l'explosion – Partie 1 : Notions fondamentales et méthodologie.

Norme NF-EN 1839 (2012). Détermination des limites d'explosivité des gaz et vapeurs.

Norme NFPA 68 (2018). Standard on explosion protection by deflagration venting.

Petit, J.M. and Poyard, J.L. (2004). Les mélanges explosifs – Partie 1 : Gaz et vapeurs. INRS, ED 911, 100.

ProSim, Software and services in process simulation (2018). Component Plus Database [Online]. Available at: www.prosim.net/en/page-simulis-20php.

Sagim, C. (1996). *Le Chlore*, 4th edition. Syndicat des halogènes et dérivés, Paris.

Strehlow, R.A., Luckritz, R.T., Adamczyk, A.A., Shimpsi, S.A. (1979). The blast wave generated by spherical flames. *Combustion and Flame*, 35, 297–310.

Triolet, J. and Sallé, B. (2009). Évaluation de la vitesse d'évaporation et de la concentration d'un composé organique volatil dans l'atmosphère d'un local de travail. INRS, ED 6058, 12.

Van Der Berg A.C. (1984). The multi-energy method – A framework for vapour cloud explosion blast prediction. Report TNO-PML. 1984-C72.

Index

A, B

D, E

F, H

I, L

M, N, O

P, R

S, T

V

Other titles from

iSTE

in

Systems and Industrial Engineering – Robotics

2022

AMARA Yacine, BEN AHMED Hamid, GABSI Mohamed
Hybrid Excited Synchronous Machines: Topologies, Design and Analysis

BOURRIÈRES Jean-Paul, PINÈDE Nathalie, TRAORÉ Mamadou Kaba, ZACHAREWICZ Grégory
From Logistic Networks to Social Networks: Similarities, Specificities, Modeling, Evaluation

DEMOLY Frédéric, ANDRÉ Jean-Claude
4D Printing 1: Between Disruptive Research and Industrial Applications
4D Printing 2: Between Science and Technology

HAJJI Rafika, JARAR OULIDI Hassane
Building Information Modeling for a Smart and Sustainable Urban Space

KROB Daniel
Model-based Systems Architecting: Using CESAM to Architect Complex Systems (Volume 3 - Systemes of Systems Complexity Set)

LOUIS Gilles
Dynamics of Aircraft Flight

2020

BRON Jean-Yves
System Requirements Engineering

KRYSINSKI TOMASZ, MALBURET FRANÇOIS
Energy and Motorization in the Automotive and Aeronautics Industries

PRINTZ Jacques
System Architecture and Complexity: Contribution of Systems of Systems to Systems Thinking

2019

ANDRÉ Jean-Claude
Industry 4.0: Paradoxes and Conflicts

BENSALAH Mounir, ELOUADI Abdelmajid, MHARZI Hassan
Railway Information Modeling RIM: The Track to Rail Modernization

BLUA Philippe, YALAOU Farouk, AMODEO Lionel, DE BLOCK Michaël, LAPLANCHE David
Hospital Logistics and e-Management: Digital Transition and Revolution

BRIFFAUT Jean-Pierre
From Complexity in the Natural Sciences to Complexity in Operations Management Systems
(Systems of Systems Complexity Set – Volume 1)

BUDINGER Marc, HAZYUK Ion, COÏC Clément
Multi-Physics Modeling of Technological Systems

FLAUS Jean-Marie
Cybersecurity of Industrial Systems

JAULIN Luc
Mobile Robotics – Second Edition Revised and Updated

KUMAR Kaushik, DAVIM Paulo J.
Optimization for Engineering Problems

TRIGEASSOU Jean-Claude, MAAMRI Nezha
Analysis, Modeling and Stability of Fractional Order Differential Systems 1: The Infinite State Approach
Analysis, Modeling and Stability of Fractional Order Differential Systems 2: The Infinite State Approach

VANDERHAEGEN Frédéric, MAAOUI Choubeila, SALLAK Mohamed, BERDJAG Denis
Automation Challenges of Socio-technical Systems

2018

BERRAH Lamia, CLIVILLÉ Vincent, FOULLOY Laurent
Industrial Objectives and Industrial Performance: Concepts and Fuzzy Handling

GONZALEZ-FELIU Jesus
Sustainable Urban Logistics: Planning and Evaluation

GROUS Ammar
Applied Mechanical Design

LEROY Alain
Production Availability and Reliability: Use in the Oil and Gas Industry

MARÉ Jean-Charles
Aerospace Actuators 3: European Commercial Aircraft and Tiltrotor Aircraft

MAXA Jean-Aimé, BEN MAHMOUD Mohamed Slim, LARRIEU Nicolas
Model-driven Development for Embedded Software: Application to Communications for Drone Swarm

MBIHI Jean
Analog Automation and Digital Feedback Control Techniques
Advanced Techniques and Technology of Computer-Aided Feedback Control

MORANA Joëlle
Logistics

SIMON Christophe, WEBER Philippe, SALLAK Mohamed
Data Uncertainty and Important Measures
(Systems Dependability Assessment Set – Volume 3)

TANIGUCHI Eiichi, THOMPSON Russell G.
City Logistics 1: New Opportunities and Challenges
City Logistics 2: Modeling and Planning Initiatives
City Logistics 3: Towards Sustainable and Liveable Cities

ZELM Martin, JAEKEL Frank-Walter, DOUMEINGTS Guy, WOLLSCHLAEGER Martin
Enterprise Interoperability: Smart Services and Business Impact of Enterprise Interoperability

2017

ANDRÉ Jean-Claude
From Additive Manufacturing to 3D/4D Printing 1: From Concepts to Achievements
From Additive Manufacturing to 3D/4D Printing 2: Current Techniques, Improvements and their Limitations
From Additive Manufacturing to 3D/4D Printing 3: Breakthrough Innovations: Programmable Material, 4D Printing and Bio-printing

ARCHIMÈDE Bernard, VALLESPIR Bruno
Enterprise Interoperability: INTEROP-PGSO Vision

CAMMAN Christelle, FIORE Claude, LIVOLSI Laurent, QUERRO Pascal
Supply Chain Management and Business Performance: The VASC Model

FEYEL Philippe
Robust Control, Optimization with Metaheuristics

MARÉ Jean-Charles
Aerospace Actuators 2: Signal-by-Wire and Power-by-Wire

POPESCU Dumitru, AMIRA Gharbi, STEFANOIU Dan, BORNE Pierre
Process Control Design for Industrial Applications

RÉVEILLAC Jean-Michel
Modeling and Simulation of Logistics Flows 1: Theory and Fundamentals
Modeling and Simulation of Logistics Flows 2: Dashboards, Traffic Planning and Management
Modeling and Simulation of Logistics Flows 3: Discrete and Continuous Flows in 2D/3D

2016

ANDRÉ Michel, SAMARAS Zissis
Energy and Environment
(Research for Innovative Transports Set - Volume 1)

AUBRY Jean-François, BRINZEI Nicolae, MAZOUNI Mohammed-Habib
Systems Dependability Assessment: Benefits of Petri Net Models (Systems Dependability Assessment Set - Volume 1)

BLANQUART Corinne, CLAUSEN Uwe, JACOB Bernard
Towards Innovative Freight and Logistics (Research for Innovative Transports Set - Volume 2)

COHEN Simon, YANNIS George
Traffic Management (Research for Innovative Transports Set - Volume 3)

MARÉ Jean-Charles
Aerospace Actuators 1: Needs, Reliability and Hydraulic Power Solutions

REZG Nidhal, HAJEJ Zied, BOSCHIAN-CAMPANER Valerio
Production and Maintenance Optimization Problems: Logistic Constraints and Leasing Warranty Services

TORRENTI Jean-Michel, LA TORRE Francesca
Materials and Infrastructures 1 (Research for Innovative Transports Set - Volume 5A)
Materials and Infrastructures 2 (Research for Innovative Transports Set - Volume 5B)

WEBER Philippe, SIMON Christophe
Benefits of Bayesian Network Models
(Systems Dependability Assessment Set – Volume 2)

YANNIS George, COHEN Simon
Traffic Safety (Research for Innovative Transports Set - Volume 4)

2015

AUBRY Jean-François, BRINZEI Nicolae
Systems Dependability Assessment: Modeling with Graphs and Finite State Automata

BOULANGER Jean-Louis
CENELEC 50128 and IEC 62279 Standards

BRIFFAUT Jean-Pierre
E-Enabled Operations Management

MISSIKOFF Michele, CANDUCCI Massimo, MAIDEN Neil
Enterprise Innovation

2014

CHETTO Maryline
Real-time Systems Scheduling
Volume 1 – Fundamentals
Volume 2 – Focuses

DAVIM J. Paulo
Machinability of Advanced Materials

ESTAMPE Dominique
Supply Chain Performance and Evaluation Models

FAVRE Bernard
Introduction to Sustainable Transports

GAUTHIER Michaël, ANDREFF Nicolas, DOMBRE Etienne
Intracorporeal Robotics: From Milliscale to Nanoscale

MICOUIN Patrice
Model Based Systems Engineering: Fundamentals and Methods

MILLOT Patrick
Designing Human–Machine Cooperation Systems

NI Zhenjiang, PACORET Céline, BENOSMAN Ryad, RÉGNIER Stéphane
Haptic Feedback Teleoperation of Optical Tweezers

OUSTALOUP Alain
Diversity and Non-integer Differentiation for System Dynamics

REZG Nidhal, DELLAGI Sofien, KHATAD Abdelhakim
Joint Optimization of Maintenance and Production Policies

STEFANOIU Dan, BORNE Pierre, POPESCU Dumitru, FILIP Florin Gh., EL KAMEL Abdelkader
Optimization in Engineering Sciences: Metaheuristics, Stochastic Methods and Decision Support

2013

ALAZARD Daniel
Reverse Engineering in Control Design

ARIOUI Hichem, NEHAOUA Lamri
Driving Simulation

CHADLI Mohammed, COPPIER Hervé
Command-control for Real-time Systems

DAAFOUZ Jamal, TARBOURIECH Sophie, SIGALOTTI Mario
Hybrid Systems with Constraints

FEYEL Philippe
Loop-shaping Robust Control

FLAUS Jean-Marie
Risk Analysis: Socio-technical and Industrial Systems

FRIBOURG Laurent, SOULAT Romain
Control of Switching Systems by Invariance Analysis: Application to Power Electronics

GROSSARD Mathieu, REGNIER Stéphane, CHAILLET Nicolas
Flexible Robotics: Applications to Multiscale Manipulations

GRUNN Emmanuel, PHAM Anh Tuan
Modeling of Complex Systems: Application to Aeronautical Dynamics

HABIB Maki K., DAVIM J. Paulo
Interdisciplinary Mechatronics: Engineering Science and Research Development

HAMMADI Slim, KSOURI Mekki
Multimodal Transport Systems

JARBOUI Bassem, SIARRY Patrick, TEGHEM Jacques
Metaheuristics for Production Scheduling

KIRILLOV Oleg N., PELINOVSKY Dmitry E.
Nonlinear Physical Systems

LE Vu Tuan Hieu, STOICA Cristina, ALAMO Teodoro,
CAMACHO Eduardo F., DUMUR Didier
Zonotopes: From Guaranteed State-estimation to Control

MACHADO Carolina, DAVIM J. Paulo
Management and Engineering Innovation

MORANA Joëlle
Sustainable Supply Chain Management

SANDOU Guillaume
Metaheuristic Optimization for the Design of Automatic Control Laws

STOICAN Florin, OLARU Sorin
Set-theoretic Fault Detection in Multisensor Systems

2012

AÏT-KADI Daoud, CHOUINARD Marc, MARCOTTE Suzanne, RIOPEL Diane
Sustainable Reverse Logistics Network: Engineering and Management

BORNE Pierre, POPESCU Dumitru, FILIP Florin G., STEFANOIU Dan
Optimization in Engineering Sciences: Exact Methods

CHADLI Mohammed, BORNE Pierre
Multiple Models Approach in Automation: Takagi-Sugeno Fuzzy Systems

DAVIM J. Paulo
Lasers in Manufacturing

DECLERCK Philippe
Discrete Event Systems in Dioid Algebra and Conventional Algebra

DOUMIATI Moustapha, CHARARA Ali, VICTORINO Alessandro, LECHNER Daniel
Vehicle Dynamics Estimation using Kalman Filtering: Experimental Validation

GUERRERO José A, LOZANO Rogelio
Flight Formation Control

HAMMADI Slim, KSOURI Mekki
Advanced Mobility and Transport Engineering

MAILLARD Pierre
Competitive Quality Strategies

MATTA Nada, VANDENBOOMGAERDE Yves, ARLAT Jean
Supervision and Safety of Complex Systems

POLER Raul *et al.*
Intelligent Non-hierarchical Manufacturing Networks

TROCCAZ Jocelyne
Medical Robotics

YALAOUI Alice, CHEHADE Hicham, YALAOUI Farouk, AMODEO Lionel
Optimization of Logistics

ZELM Martin *et al.*
Enterprise Interoperability –I-EASA12 Proceedings

2011

CANTOT Pascal, LUZEAUX Dominique
Simulation and Modeling of Systems of Systems

DAVIM J. Paulo
Mechatronics

DAVIM J. Paulo
Wood Machining

GROUS Ammar
Applied Metrology for Manufacturing Engineering

KOLSKI Christophe
Human–Computer Interactions in Transport

LUZEAUX Dominique, RUAULT Jean-René, WIPPLER Jean-Luc
Complex Systems and Systems of Systems Engineering

ZELM Martin, *et al.*
Enterprise Interoperability: IWEI2011 Proceedings

2010

BOTTA-GENOULAZ Valérie, CAMPAGNE Jean-Pierre, LLERENA Daniel, PELLEGRIN Claude
Supply Chain Performance / Collaboration, Alignement and Coordination

BOURLÈS Henri, GODFREY K.C. Kwan
Linear Systems

BOURRIÈRES Jean-Paul
Proceedings of CEISIE'09

CHAILLET Nicolas, REGNIER Stéphane
Microrobotics for Micromanipulation

DAVIM J. Paulo
Sustainable Manufacturing

GIORDANO Max, MATHIEU Luc, VILLENEUVE François
Product Life-Cycle Management / Geometric Variations

LOZANO Rogelio
Unmanned Aerial Vehicles / Embedded Control

LUZEAUX Dominique, RUAULT Jean-René
Systems of Systems

VILLENEUVE François, MATHIEU Luc
Geometric Tolerancing of Products

2009

DIAZ Michel
Petri Nets / Fundamental Models, Verification and Applications

OZEL Tugrul, DAVIM J. Paulo
Intelligent Machining

PITRAT Jacques
Artificial Beings

2008

ARTIGUES Christian, DEMASSEY Sophie, NERON Emmanuel
Resources–Constrained Project Scheduling

BILLAUT Jean-Charles, MOUKRIM Aziz, SANLAVILLE Eric
Flexibility and Robustness in Scheduling

DOCHAIN Denis
Bioprocess Control

LOPEZ Pierre, ROUBELLAT François
Production Scheduling

THIERRY Caroline, THOMAS André, BEL Gérard
Supply Chain Simulation and Management

2007

DE LARMINAT Philippe
Analysis and Control of Linear Systems

DOMBRE Etienne, KHALIL Wisama
Robot Manipulators

Printed and bound by CPI Group (UK) Ltd, Croydon, CR0 4YY

16/09/2026

14963081-0007